the

the Universal Computer

The Road from Leibniz to Turing

Third Edition

Martin Davis

CRC Press is an imprint of the
Taylor & Francis Group, an **informa** business

CRC Press
Taylor & Francis Group
6000 Broken Sound Parkway NW, Suite 300
Boca Raton, FL 33487-2742

Printed by Ashford Colour Press Ltd.
Version Date: 20180109

International Standard Book Number-13: 978-1-1385-0208-6 (Paperback)
International Standard Book Number-13: 978-0-8153-8402-1 (Hardback)

Visit the Taylor & Francis Web site at
http://www.taylorandfrancis.com

and the CRC Press Web site at
http://www.crcpress.com

To Virginia, my life's companion

Contents

Preface to Third Edition

For this edition, I have had the opportunity to write about deep learning algorithms and the astonishing AlphaGo program that defeats human masters of the ancient game of Go. I tried to disentangle what is known about the relationship between Georg Cantor and Leopold Kronecker from the widespread myths about their relationship. In addition, I have detailed the drama of ideas in Princeton during the 1930s involving Alonzo Church, his student Stephen Kleene, and Kurt Gödel, before Alan Turing's arrival.

I am grateful to Susan Dickie, Harold Edwards, Dana Scott and François Treves who helped me in various ways. I'm particularly grateful to Thore Graepelof DeepMind for patiently explaining AlphaGo to me. Finally, I want to thank Sarfraz Khan, an editor who understood what I was trying to do and shared my enthusiasm for the project.

Martin Davis
Berkeley, 2017

Preface to Second Edition

Alan Turing was born on June 23, 1912. The year 2012, the hundredth anniversary of his birth, is providing the occasion for events and publications reflecting on and celebrating his achievements. I am delighted that this updated version of my book will be part of the excitement, and I am very grateful to Klaus Peters and Alice Peters for their help in making it happen. For this edition, I've tidied up some loose ends and brought a few things up to date, even commenting on IBM's achievement in fielding its Watson computer as a successful contestant on the popular television quiz program *Jeopardy*.

The "universality" of the computers with which we interact today is evident in the myriad unrelated tasks for which the computers that are on our desks or our laps are used, but also in the hidden computers that are embedded in many other devices. If our cameras are computers with lenses and our telephones are computers with microphones and earphones, it could almost be said that the hybrid automobile I drive is a computer with four wheels.

Today it is widely recognized that this universality is an application of a fundamental insight from an article by Turing published in a mathematical journal in 1936. When I began researching these matters in the 1980s, much controversy surrounded credit for the first "stored program" electronic computers, but Turing's name was never mentioned. The argument was whether the credit was due to the mathematician John von Neumann or to the engineers John Presper Eckert and John Mauchly. David Leavitt kindly suggested that I am responsible for the recognition of Turing's role. While an article I wrote may have had some influence, probably the publication of some of Turing's previously unavailable work from the 1940s together with the availability of information about his secret work towards decrypting enemy military communications during the Second World War was more important.[1]

This is a book of stories about seven remarkable people, their ideas and discoveries, and their fascinating lives. They investigated the why and how of logical reasoning. They forced the exhilaration and pitfalls of trying to come to grips with the infinite. Their heroic efforts to buttress the claims of rationality encountered unforeseen obstacles. Finally, Alan Turing's radically new understanding of the nature of algorithmic processes and its potential to make a single "all-purpose" machine that could be programmed to carry out almost any process was a by-product of this tumultuous development. I had great fun writing this book, and I hope that you will have fun reading it.

Berkeley, June 30, 2011

Preface

This book is about the underlying concepts on which our modern computers are based and about the people who developed these concepts. In the spring of 1951, shortly after completing my doctorate in mathematical logic at Princeton University where Alan Turing worked a decade earlier, I was teaching a course at the University of Illinois based on his ideas. A young mathematician who attended my lectures called my attention to a pair of machines being constructed across the street from my classroom that he insisted were embodiments of Turing's conception. It was not long before I found myself writing software for early computers. My professional career spanning half a century revolved around this relationship between the abstract logical concepts underlying modern computers and their physical realization.

As computers evolved from the room-filling behemoths of the 1950s to the small powerful machines of today performing a bewildering variety of tasks, their underlying logic has remained the same. These logical concepts developed from the work of a number of gifted thinkers over centuries. In this book I tell the stories of the lives of these people and explain some of their thoughts. The stories are fascinating. My hope is that readers will not only enjoy them, but will come away with a better sense of what goes on inside their computers and with enhanced respect for the value of abstract thought.

In developing this book I benefited from help of various kinds. The John Simon Guggenheim Memorial Foundation provided welcome financial support during the early stages of the studies that led to this book. Patricia Blanchette, Michael Friedman, Andrew Hodges, Lothar Kreiser, and Benson Mates generously shared their expert knowledge with me. Tony Sale kindly acted as my guide to Bletchley Park where Turing played an important part in the decoding of secret German military communications during World War II. Eloise Segal, who alas did not live to see the book completed, was a devoted reader who helped me avoid expository pitfalls. My wife, Virginia, stubbornly refused to let me be obscure. Sherman Stein read the manuscript carefully and suggested many improvements while saving me from a number of errors. I benefited from help with translations by Egon Börger, William Craig, Michael Richter, Alexis Manaster Ramer, Wilfried Sieg, and Francois Treves. Other readers who provided useful comments were Harold Davis, Nathan Davis, Jack Feldman, Meyer Garber, Dick and Peggy Kuhns, and Alberto Policriti. My editor, Ed Barber at W.W. Norton, generously shared his knowledge of English prose style and is responsible for many improvements. Harold Rabinowitz introduced me to my agent, Alex Hoyt, who has been unfailingly helpful. Of course this long list of names is meant only to express gratitude and not to absolve myself of responsibility for the book's shortcomings. I would be grateful for comments or corrections from readers sent to me at: davis@eipye.com.

Martin Davis
Berkeley, January 2, 2000

Introduction

If it should turn out that the basic logics of a machine designed for the numerical solution of differential equations coincide with the logics of a machine intended to make bills for a department store, I would regard this as the most amazing coincidence I have ever encountered.

—Howard Aiken 1956[1]

Let us now return to the analogy of the theoretical computing machines ... It can be shown that a single special machine of that type can be made to do the work of all. It could in fact be made to work as a model of any other machine. ***The special machine may be called the universal machine*** *...*

—Alan Turing 1947[2]

In the fall of 1945, as the ENIAC, a gigantic calculating engine containing thousands of vacuum tubes, neared completion at the Moore School of Electrical Engineering in Philadelphia, a group of experts met regularly to discuss the design of its proposed successor, the EDVAC. As the weeks went by, the meetings grew increasingly acrimonious, with the experts finding themselves divided into two groups they dubbed the "engineers" and the "logicians." John Presper Eckert, leader of the "engineers," was justly proud of his accomplishment with the ENIAC. It had been thought impossible for 15,000 hot vacuum tubes to work together long enough without any of them failing, for anything useful to be accomplished. Nevertheless, by using careful conservative design principles, Eckert had succeeded brilliantly in accomplishing this feat. Things came to a head when, much to Eckert's displeasure, the group's leading "logician," the eminent mathematician John von Neumann, circulated, under his own name, a draft report on the proposed EDVAC that, paying little attention to engineering details, set forth the fundamental *logical* computer design known to this day as the von Neumann architecture.

Although an engineering tour de force, the ENIAC was a logical mess. It was von Neumann's expertise as a logician and what he had learned from the English logician Alan Turing that enabled him to understand the fundamental fact that a computing machine is a logic machine. In its circuits are embodied the distilled insights of a remarkable collection of logicians, developed over centuries. Nowadays, when computer technology is advancing with such breathtaking rapidity, as we admire the truly remarkable accomplishments of the engineers, it is all too easy to overlook the logicians whose ideas made it all possible. This book tells their story.

CHAPTER 1

Leibniz's Dream

Situated southeast of the German city of Hanover, the ore-rich veins of the Harz mountain region had been mined since the middle of the tenth century. Because the deeper parts tended to fill with water, they could only be mined so long as pumps kept the water at bay. During the seventeenth century water wheels powered these pumps. Unfortunately, this meant that the lucrative mining operations had to shut down during the cold mountain winter season when the streams were frozen.

During the years 1680–1685, the Harz mountain mining managers were in frequent conflict with a most unlikely miner. G. W. Leibniz, then in his middle thirties, was there to introduce windmills as an additional energy source to enable all-season operation of the mines. At this point in his life, Leibniz had already accomplished a lot. Not only had he made major discoveries in mathematics, he had also acquired fame as a jurist, and had written extensively on philosophical and theological issues. He had even undertaken a diplomatic mission to the court of Louis XIV in an attempt to convince the French "Sun King" of the advantages of conducting a military campaign in Egypt (instead of against Holland and German territories).[1]

Some 70 years earlier, Cervantes had written of the misadventures of a melancholy Spaniard with windmills. Unlike Don Quixote, Leibniz was incurably optimistic. To those who reacted bitterly to the widespread misery in the world, Leibniz responded that God, from His omniscient view of all possible worlds, had unerringly created the best that could be constructed, that all the evil elements of our world were balanced by good in an optimal manner.*

But Leibniz's involvement with the Harz Mountain mining project ultimately proved to be a fiasco. In his optimism, he had not foreseen the natural hostility of the expert mining engineers towards a novice proposing to teach them their trade. Nor had he allowed for the inevitable break-in period a novel piece of machinery requires or for the unreliability of the winds. But his most incredible piece of optimism was with respect to what he had imagined he would be able to accomplish with the proceeds he had expected from the project.

*Voltaire's Dr. Pangloss in Voltaire's *Candide* was a sendup of this Leibnizian doctrine.

Gottfried Wilhelm Leibniz

Leibniz had a vision of amazing scope and grandeur. The notation he had developed for the differential and integral calculus, the notation still used today, made it easy to do complicated calculations with little thought. It was as though the notation did the work.

In Leibniz's vision, something similar could be done for the whole scope of human knowledge. He dreamt of an encyclopedic compilation, of a universal artificial mathematical language in which each facet of knowledge could be expressed, of calculational rules which would reveal all the logical interrelationships among these propositions. Finally, he dreamed of machines capable of carrying out calculations, freeing the mind for creative thought. Even with his optimism, Leibniz knew that the task of transforming this dream to reality was not something he could accomplish alone. But he did believe that a small number of capable people working together in a scientific academy could accomplish much of the design in a few years. It was to fund such an academy that Leibniz embarked on his Harz Mountain project.

Leibniz's Wonderful Idea

Leibniz was born in Leipzig in 1646 into a Germany divided into something like 1,000 separate, semiautonomous political units, and devastated by almost three decades of war. The Thirty Years War, which didn't end until 1648, was fought mainly on German soil, although all of the major European powers had participated. Leibniz's father, a professor of philosophy at the University of Leipzig, died when the child was only six. Over the opposition of his teachers, Leibniz gained access to his father's library at the age of eight, and soon became a fluent reader of Latin.

Leibniz, destined to become one of the greatest mathematicians of all time, got his first introduction to mathematical ideas from teachers who had no inkling of the exciting work elsewhere in Europe that was revolutionizing mathematics. In the Germany of that day, even the elementary geometry of Euclid was an advanced subject, studied only at the university level. However, in his early teens, his school teachers did introduce Leibniz to the system of logic that Aristotle had developed two millennia earlier, and this was the subject that aroused his mathematical talent and passion.

Fascinated by the Aristotelian division of concepts into fixed "categories," Leibniz thought of what he came to call his "wonderful idea": he would seek a special "alphabet" whose elements represented not sounds, but concepts. A language based on such an alphabet should make it possible to determine by symbolic calculation which sentences written in the language were true and what logical relationships existed among them. Leibniz remained under Aristotle's spell and held fast to this vision for the rest of his life.

Indeed, for his bachelor's degree at Leipzig, Leibniz wrote a thesis on Aristotelian metaphysics. His master's thesis at the same university dealt with the relationship between philosophy and law. Evidently attracted to legal studies, Leibniz obtained a second bachelor's degree, this time in law, writing a thesis emphasizing the use of systematic logic in dealing with the law. Leibniz's first real contribution to mathematics developed out of his *Habilitationsschrift* (in Germany, a kind of second doctoral dissertation) in philosophy also at Leipzig: As a first step towards his "wonderful idea" of an alphabet of concepts, Leibniz foresaw the need to be able to count the various ways of combining such concepts. This led him to a systematic study of the problem of counting complex arrangements of basic elements, first in his *Habilitationsschrift* and then in his more extensive monograph *Dissertatio de Arte Combinatoria.*[2]

Continuing his legal studies, Leibniz presented a dissertation for a doctorate in law at the University of Leipzig. The subject, so typical for Leibniz, was the use of reason to resolve cases in law thought too difficult for resolution by the normal methods. For reasons that are not clear the Leipzig faculty refused to accept the dissertation, so Leibniz presented it instead at the University of Altdorf, near Nuremberg where it was received with acclaim. At the age of 21, his formal education completed, Leibniz faced the common problem of the newly graduated: how to develop a career.

Paris

Not being interested in a career as a university professor in Germany, Leibniz pursued his only real alternative: to find a wealthy noble patron. Baron Johann von Boineburg, nephew of the Elector of Mainz, was quite willing to play this role. In Mainz, Leibniz worked on a project to update the legal system based on Roman civil law, was appointed a judge at the High Court of Appeal, and tried his hand at diplomatic intrigue. This last included an abortive attempt to influence the selection of a new king for Poland and a mission to the court of Louis XIV.

The Thirty Years War had left France as the "superpower" on the European continent. Mainz, situated on the banks of the Rhine, had known military occupation during the war. So, the burghers of Mainz understood very well the importance of forestalling hostile military action, and therefore, of good relations with France. It was in this context that Boineburg and Leibniz concocted the scheme, already mentioned, to convince Louis XIV and his advisers of the great advantages of making Egypt the object of their military endeavors. The most important historical effect of this proposition—essentially the same proposition that led Napoleon to a military disaster over a century later—was that it brought Leibniz to Paris.

Leibniz arrived in Paris in 1672 to press the Egyptian scheme and to help untangle some of Boineburg's financial affairs. Before the end of the year disaster struck: the news came that Boineburg had died of a stroke. Although he continued to perform some services for the Boineburg family, Leibniz was left without a reliable source of income. Nevertheless he managed to remain in Paris for another four extremely productive years that included two brief visits to London.[3] On the first visit in 1673, he was unanimously elected to the Royal Society of London based on his model of a calculating machine capable of carrying out the four basic operations of arithmetic. Although Pascal had designed a machine that could add and subtract, Leibniz's was the first that could multiply and divide as well.* Leibniz's machine incorporated an ingenious gadget that became known as a "Leibniz wheel." Calculating machines continued to be built incorporating this device well into the twentieth century. About his machine, Leibniz wrote:

> And now that we may give final praise to the machine we may say that it will be desirable to all who are engaged in computations which, it is well known, are the managers of financial affairs, the administrators of others' estates, merchants, surveyors, geographers, navigators, astronomers ... But limiting ourselves to scientific uses, the old geometric and astronomic tables could be corrected and new ones constructed by the help of which we could measure all kinds of curves and figures ... it will pay to extend as far as possible the major Pythagorean tables; the table of squares, cubes, and other powers; and the tables of combinations, variations, and progressions of all kinds, ... Also the astronomers surely will not have to continue to exercise the patience which is required for computation. ... For it is unworthy of excellent men to lose hours like slaves in the labor of calculation which could safely be relegated to anyone else if the machine were used.[4]

The machine Leibniz was "praising" was limited to ordinary arithmetic. But Leibniz grasped the broader significance of mechanizing calculation. In 1674 he described a machine that could solve algebraic equations. A year later, he wrote comparing logical reasoning to a mechanism, thus pointing to the goal of reducing reasoning to a kind of calculation and of ultimately building a machine capable of carrying out such calculations.[5]

A crucial event for Leibniz, then 26, was meeting the great Dutch scientist Christiaan Huygens then living in Paris. The 43-year-old Huygens had already invented the pendulum clock and discovered the rings of Saturn.

*Blaise Pascal, born on June 19, 1623, at Clermont-Ferrand, France, one of the founders of the mathematical theory of probability, was a prolific mathematician, physicist, and religious philosopher. His calculating machine, designed and built circa 1643, brought him considerable fame. He died in 1662.

What was perhaps to be his most important contribution, the wave theory of light, was still to come. His conception—that light was fundamentally to be viewed like the waves spreading across a pond when a pebble is tossed into it—directly contradicted the great Newton's account of light as consisting of a stream of discrete bullet-like particles.* Huygens gave Leibniz a reading list enabling the younger man to quickly overcome his lack of knowledge of current mathematical research. Soon Leibniz was making fundamental contributions.

The explosion of mathematical research in the seventeenth century had been fueled by two crucial developments:

1. The technique of dealing with algebraic expressions (what is generally called "high-school algebra") had been systematized and became essentially the powerful technique we still use today.

2. Descartes and Fermat had shown how, by representing points by pairs of numbers, geometry could be reduced to algebra.

Various mathematicians were using this new power to solve problems that would not previously have been accessible. Much of this work involved what nowadays are called *limit processes*. Using limits means solving a problem by using approximations to the required answer that get systematically closer and closer to that answer. The idea was not to be satisfied with any particular approximation, but rather, by "going to the limit," to obtain an *exact* solution.

An example that may help to clarify this concept is one of Leibniz's own early results, one of which he was quite proud. This was the equation:

$$\frac{\pi}{4} = 1 - \frac{1}{3} + \frac{1}{5} - \frac{1}{7} + \frac{1}{9} - \frac{1}{11} + \cdots$$

On the left side of the equals sign is the familiar number π that occurs in the formulas for the circumference and the area of a circle.* On the right side is what is called an *infinite series*; the individual numbers alternately added and subtracted are called the *terms* of the series. The dots ... mean that it continues indefinitely. The full infinite pattern consists of fractions, with 1 as numerator and the successive odd numbers as denominators, being alternately added and subtracted, and is intended to be clear from the finite part shown: after subtracting $\frac{1}{11}$, add $\frac{1}{13}$, then subtract $\frac{1}{15}$, etc. But can one actually perform an infinite number of additions and subtractions? Not really. But, starting at the beginning and breaking off at any point, an

*Although Huygens's view came to be generally accepted, the coming of quantum physics in the twentieth century made it clear that both Newton and Huygens had been right; each grasped an essential characteristic of light.

*The number $\frac{\pi}{4}$ is in fact the area of a circle whose radius is $\frac{1}{2}$.

Number of terms	Sum correct to eight decimal places
10	0.76045990
100	0.78289823
1,000	0.78514816
10,000	0.78537316
100,000	0.78539566
1,000,000	0.78539792
10,000,000	0.78539816

Table of approximations to Leibniz's series

approximation to a "true" answer is obtained, and that approximation gets better and better as more terms are included. In fact, the approximation can be made as accurate as one wishes by including enough terms. In the table on page 7, it is shown how this works out for Leibniz's series. When including 10,000,000 terms, a value is obtained that agrees with the true value of $\frac{\pi}{4}$, namely 0.7853981634..., to eight places.†

Leibniz's series is so striking because it connects the number π, and therefore the area of a circle, with the succession of odd numbers in a particularly simple way. It is an example of one kind of problem that could be solved using limit processes, that of finding areas of figures with curved boundaries.

Another kind of problem susceptible to attack using limits was finding exact rates of change, such as the varying speed of a moving body. During the last months of 1675, towards the end of his stay in Paris, Leibniz made a number of conceptual and computational breakthroughs in the use of limit processes that, taken together, are called his "invention of the calculus":

1. Leibniz saw that the problems of finding areas and calculating rates of change were paradigmatic, in the sense that many different kinds of problems were reducible to one or the other of these two types.*

2. He also perceived that the mathematical operations required in calculating the solutions to problems of these two types were in fact *inverse* to each other in much the same sense that the operations of addition and subtraction (or multiplication and division) are inverse to one

†I used my PC to obtain the table of approximations to $\frac{\pi}{4}$ from Leibniz's series. A short Pascal program I wrote for the purpose runs for less than a second on a contemporary PC.

*Thus, finding volumes and centers of gravity are problems of the first kind, and computing accelerations and (in economic theory) marginal elasticity are problems of the second type.

another. Nowadays these operations are called *integration* and *differentiation*, respectively, and the fact that they are inverse is called, in the textbooks, the "fundamental theorem of the calculus."

3. Leibniz developed an appropriate symbolism (the notation still in use today) for these operations, $\int$ for integration and d for differentiation.† Finally he found the mathematical rules needed for carrying out the integrations and differentiations that occurred in practice.

Taken together these discoveries transformed the use of limit processes, from an exotic method accessible only to a handful of specialists, into a straightforward technique that could be taught in textbooks to many thousands of people.[6] Most important for the purposes of this book, his success convinced Leibniz of the critical importance of choosing appropriate symbols and finding the rules governing their manipulation. The symbols $\int$ and d did not represent meaningless sounds like the letters of a phonetic alphabet; they stood for concepts and thus provided a model for Leibniz's boyhood "wonderful idea" of an alphabet representing *all* fundamental concepts.

Much has been written about the entirely independent development of the calculus by Newton and by Leibniz, and about the bitter accusations of plagiarism tossed back and forth across the English Channel before the foolishness of such charges was finally understood by all. It is the great superiority of Leibniz's notation that is significant for our story.[7] A key technique used in integration (called in the textbooks, the method of "substitution") is virtually automatic in Leibniz's notation, but relatively complicated in Newton's. It has even been alleged that slavish devotion to their national hero's methods caused the English followers of Newton to lag far behind their continental contemporaries in developing the mathematical perspectives that the calculus had uncovered.

Like so many who have tasted the special quality of life in Paris, Leibniz wanted very much to remain there as long as he could. He attempted to maintain his Mainz connections while continuing to live and work in Paris. But it soon became clear that, so long as he remained in Paris no funds from Mainz would be forthcoming.

Meanwhile an offer of a position arrived from the Dukedom of Hanover, one of the multitude of principalities of which seventeenth century Germany was composed. Although Duke Johann Friedrich had some genuine interest in intellectual matters, and the offer gave some promise of financial security, Leibniz was not eager to live in Hanover. After delaying as long as he could, Leibniz accepted the offer early in 1675.

†The symbol for integration $\int$ is actually a modified "S" intending to suggest "sum," and the symbol "*d*" is likewise intended to suggest the idea of "difference."

In his letter of acceptance, he asked for the "freedom to pursue his own studies in arts and sciences for the benefit of mankind."[8] In no hurry to leave Paris, he stayed until the fall of 1676, departing only when it became clear that no position in Paris would be forthcoming and that the Duke would accept no further delay. Leibniz was to spend the rest of his life in the service of the Dukes of Hanover.

Hanover

Leibniz apparently understood perfectly well that despite his request for "freedom to pursue his own studies in arts and sciences," success in his new position would require him to do work that his patron would find useful and practical. He undertook to upgrade the ducal library and proposed various ideas for improving public administration and agriculture. Soon thereafter, he began promoting his ill-fated project to use windmills for improving the Harz Mountain mining operations. In 1680, only a year after the Harz project with Leibniz in charge had finally been approved, his position was suddenly endangered by the duke's sudden death.

It now became necessary to convince the new duke, Ernst August, to continue to found Leibniz's position and to support the Harz Mountain project. The new duke was a practical man. Unlike his predecessor, he wasn't willing to spend much on the library. Leibniz soon learned not to involve Ernst August in scholarly discussions.

To help cement his position, he offered to write a short history of the duke's family. When the duke finally closed down the Harz project five years later, Leibniz proposed a more elaborate version: if a few gaps were filled, the family tree could be traced back to the year 600. The duke evidently regarded this as a most appropriate way to employ one of the greatest thinkers of all time. Leibniz was granted a regular salary, a personal secretary, and travel funds for searching out genealogical information. Most likely, the optimistic Leibniz hardly imagined that he would find himself chained to this project for the remaining three decades of his life. Georg Ludwig, who succeeded Ernst August on his death in 1698, was especially adamant in nagging Leibniz to get on with the family history.

If Leibniz had any pupils in Hanover, they were women, for he shared none of the common prejudices concerning the intellectual capabilities of the female sex. Duchess Sophie, the talented wife of Ernst August, and Leibniz had frequent conversations about philosophical matters and carried on an extensive correspondence when Leibniz was away from Hanover. She made sure also that her daughter Sophie Charlotte, who was to become Queen of Prussia, also had the benefit of Leibniz's teachings. Sophie Charlotte, not content simply to receive Leibniz's wisdom, energetically raised questions that helped Leibniz to clarify his ideas. As the contemporary Leibniz scholar Benson Mates explains:

> For most of Leibniz's life, these women were his principal advocates at the courts in Hanover and Berlin. Sophie Charlotte's sudden death in 1705 devastated him; it was such an obvious loss to him that he even received formal expressions of sympathy from the emissaries of foreign governments; and when Duchess Sophie ... died in 1714, his ability to obtain support for anything other than continuing the Brunswick history came to an end.[9]

The history project did provide Leibniz with an excuse to travel, and he made use of this freedom to an extent that vexed his noble patrons. Of course Leibniz took full advantage of the possibilities of developing and maintaining scholarly contacts. In Berlin he even was able to found a Society of Science, later institutionalized as an academy. His extensive correspondence continued to span the full variety of his interests.

Leibniz seemed never to tire of explaining that, since God had done as well as was possible in creating the world, there must be a *pre-established harmony* between what existed and what was possible and that there was a *sufficient reason* (whether or not we could find it) for every single thing in the world.

In the realm of diplomacy, Leibniz had two pet projects: one was to reunite the various branches of the Christian church; the other, which actually succeeded, was to obtain for the Dukes of Hanover the succession to the British throne. But when Georg Ludwig actually did become George I of England only two years before Leibniz's death in 1716, he brusquely rejected his employee's request for permission to leave the Hanovarian backwater for London with his patron, ordering him to hurry up and finish the family history.

The Universal Characteristic

But what of the "wonderful idea" of Leibniz's youth, his grand dream to find a true alphabet of human thought and the appropriate calculational tools for manipulating these symbols? Although he had resigned himself to the fact that unaided he could never accomplish such a thing, he never lost sight of this goal, thinking and writing about it throughout his life. It was clear that the special characters used in arithmetic and algebra, the symbols used in chemistry and astronomy, and the symbols he introduced for the differential and integral calculus provided a paradigm showing how crucial a truly appropriate symbolism could be.

Leibniz referred to such a system of characters as a *characteristic*. Unlike the alphabetic symbols which had no meaning, the examples just mentioned were, for him, a *real characteristic* in which each symbol represented some definite idea in a natural and appropriate way. What was needed, Leibniz maintained, was a *universal characteristic,* a system of symbols that was not only *real,* but which also encompassed the full scope of human thought.

In a letter explaining this to the mathematician G. F. A. l'Hôspital, Leibniz wrote: "Part of the secret of" algebra "consists of the characteristic, that is to say of the art of properly using" the symbolic expressions. This care for proper use of symbols was to be the "thread of Ariadne" that would guide the scholar in creating his characteristic.

As the early twentieth century logician and Leibniz scholar Louis Couturat explained:

> ...it is algebraic notation that incarnates, so to speak, the ideal of the characteristic and which is to serve as a model. It is also the example of algebra that Leibniz cites consistently to show how a system of properly chosen symbols is useful and indeed indispensible for deductive thought.[10]

Perhaps the most enthusiastic explanation of his proposed characteristic was in another letter, this one to Jean Galloys with whom Leibniz had extensive correspondence:

> I am convinced more and more of the utility and reality of this general science, and I see that very few people have understood its extent. ... This characteristic consists of a certain script or language ...that perfectly represents the relationships between our thoughts. The characters would be quite different from what has been imagined up to now. Because one has forgotten the principle that the characters of this script should serve invention and judgment as in algebra and arithmetic. This script will have great advantages; among others, there is one that seems particularly important to me. This is that it will be impossible to write, using these characters, chimerical notions (*chimères*) such as suggest themselves to us. An ignoramus will not be able to use it, or, in striving to do so, he himself will become erudite.[11]

In the letter to Galloys quoted above Leibniz refers to arithmetic as well as algebra as showing the importance of an appropriate symbolism. He had in mind in particular the advantage of the Arabic system of notation that we still use today based on the digits 0 to 9 over previous systems (like the Roman numerals) for ordinary calculation. When Leibniz discovered binary notation, in which any number can be written using only the digits 0 and 1, he was impressed by the simplicity of this system. He believed that it would be useful in laying bare properties of numbers that otherwise would be hidden. Although this belief turned out to be unjustified, this interest on Leibniz's part is remarkable in the light of the importance of this binary notation in connection with modern computers.

Leibniz saw his grand program as consisting of three major components. First, before the appropriate symbols could be selected, it would be necessary to create a compendium or *encyclopedia* encompassing the full extent of human knowledge. He maintained that once having accomplished this, it should prove feasible to select the key underlying notions and to provide appropriate symbols for each of them. Finally, the rules of deduction could then be reduced to manipulations of these symbols, via what Leibniz called a *calculus ratiocinator*, what nowadays might be called a symbolic logic.

To a present-day reader, it is hardly surprising that Leibniz did not feel able to accomplish such a program on his own, especially given the constant pressure he was under to produce the family history that his patron regarded as his principal task. It is difficult to understand how Leibniz could have seriously believed that the universe we inhabit, in all of its complexity, could be reduced to a single symbolic calculus.

We can only hope to begin to comprehend the matter by attempting to see the world through the eyes of Leibniz. For him nothing, absolutely nothing, about the world was in any way undetermined or accidental. Everything was in fact entirely determined by the plan, clear in the mind of God, by means of which He had created the best world that could be created. Hence, for Leibniz, all aspects of the world, natural and supernatural, were connected by links one could hope to discover by rational means. Only from this perspective can we understand how, in a famous passage, Leibniz could write of serious "men of good will" sitting around a table to solve some critical problem. After writing out the problem in Leibniz's projected language, his "universal characteristic," it would be time to say "Let us calculate!" Out would come the pens and a solution would be found whose correctness would necessarily be accepted by all.[12]

Leibniz wrote with enthusiasm about the importance of producing the *calculus ratiocinator*, the algebra of logic, that would presumably be needed to carry out these calculations:

> For if praise is given to the men who have determined the number of regular solids—which is of no use, except insofar as it is pleasant to contemplate—and if it is thought to be an exercise worthy of a mathematical genius to have brought to light the more elegant properties of a conchoid or cissoid, or some other figure which rarely has any use, how much better will it be to bring under mathematical laws human reasoning, which is the most excellent and useful thing we have.[13]

Unlike the universal characteristic concerning which Leibniz wrote with such passion and conviction, but produced little in the way of specifics, he

---

DEFINITION 3. A is in L, or L contains A, is the same as to say that L can be made to coincide with a plurality of terms taken together of which A is one. $B \oplus N = L$ signifies that B is in L and that B and N together compose or constitute L. The same thing holds for a larger number of terms.

AXIOM 1. $B \oplus N = N \oplus B$.

POSTULATE. Any plurality of terms, as A and B, can be added to compose a single term $A \oplus B$.

AXIOM 2. $A \oplus A = A$.

PROPOSITION 5. *If A is in B and $A = C$, then C is in B.* For in the proposition A *is in* B the substitution of A for B gives C *is in* B.

PROPOSITION 6. *If C is in B and $A = B$ then C is in A.* For in the proposition C *is in* B the substitution of A for B gives C *is in* A.

PROPOSITION 7. A *is in* A. For A is in $A \oplus A$ (by Definition 3). Therefore (by Proposition 6) A is in A.

. . . .

PROPOSITION 20. *If A is in M and B is in N, then $A \oplus B$ is in $M \oplus N$.*

Sample from one of Leibniz's logical Calculi

did make a number of attempts to produce a *calculus ratiocinator*. Part of his most polished effort in this direction is shown in the above illustration.[14] A good century and a half ahead of his time, Leibniz proposed an algebra of logic that would specify the rules for manipulating logical concepts in the manner in which ordinary algebra specifies the rules for manipulating numbers. He introduced a special new symbol $\oplus$ to represent the combining of arbitrary "pluralities of terms." The idea was something like the combining of two collections of things into a single collection containing all of the items in either one. The plus sign encourages us to think of this operation as being like ordinary addition, but the circle around it warns us that it is not exactly like ordinary addition because it is not numbers being added. Some of his algebraic rules are also to be found in high-school algebra textbooks: to some extent the same rules work for logical concepts as for numbers.

But there's more to the story. There are also rules that are very different from those for numbers. The most striking rule of this latter kind, one that in a somewhat different context George Boole was to make the cornerstone of his algebra of logic, is Leibniz's Axiom 2, $A \oplus A = A$, which expresses the fact that combining a "plurality of terms" with itself will yield nothing new: evidently combining all the things belonging to a given collection with that same collection of things, will just produce that same collection, all over again. Of course addition of numbers is quite different: $2 + 2 = 4$ not 2.

In the next chapter, we will see how George Boole, presumably ignorant of Leibniz's efforts, produced a serviceable symbolic logic along the lines that Leibniz had pioneered. Boole's logic subsumed the logic Aristotle had introduced 2000 years earlier, but it was only with the work of Gottlob Frege well into the nineteenth century, that the serious limitations shared by the logical systems of Aristotle and of Boole were really overcome.[15]

Despite Leibniz's voluminous correspondence, we have little idea of what he was like as a person. One biographer claims to see in the few portraits of Leibniz we possess, the image of a tired, unhappy, pessimistic man, in contradiction to his optimistic philosophy.[16] Others have remarked that he liked to give cakes to his neighbors' children. Apparently, he proposed marriage when he was 50, but thought better of it when the lady hesitated.[17] We have the picture of Leibniz spending long days and often entire nights seated at his desk managing his enormous correspondence with remarkable punctuality, his meals brought to him from an inn by his servants. What is clear is that he was indefatigable in his work.*

It is tempting to indulge in a bit of "what if?" What if Leibniz had not been shackled to his patrons' family history, and was free to devote more time to his *calculus rationcinator*? Might he not have accomplished what Boole was only to do so much later? But of course, such speculation is useless. What Leibniz has left us is his dream, but even this dream can fill us with admiration for the power of human speculative thought and can serve as a yardstick for judging later developments.

*In part, this picture comes from the 1951 biography completed by Professor Kurt Huber in prison while awaiting execution by the Nazis. He had supported the efforts of his students at the University of Munich who formed the "White Rose" underground group and were decapitated for distributing anti-Nazi leaflets. There are today a number of streets in Germany named for him including a Professor Huber Platz at the University of Munich. (I am indebted to Benson Mates for this information about Professor Huber's heroic role.)

CHAPTER 2

Boole Turns Logic into Algebra

George Boole's Hard Life

The beautiful and intelligent Princess Caroline von Ensbach, one day to be Queen of England as the wife of George II, met Leibniz in Berlin in 1704 when she was 18. After she went to England with the court, their friendship continued by correspondence. She tried to persuade her father-in-law, then George I of England, to bring Leibniz to England, but as we have seen, the king insisted that Leibniz remain in Germany to complete the Hanoverian family history.

Caroline found herself entangled in the continuing dispute between Leibniz and Newton and his followers, each side accusing the other of plagiarism over the invention of the calculus. She tried to convince Leibniz that the issue was of no great importance, but he was having none of it. Indeed, Leibniz sought her support before the king for his desire to be appointed "Historiographer of England" to match Newton's position as "Master of the Mint," asserting that only in this way could the honor of Germany vis a vis England be maintained.

Leibniz wrote Caroline that when Newton held that a grain of sand exerted a gravitational force on the distant sun without any evident means by which such a force could be transmitted, he was in effect calling on miraculous means to explain a natural phenomenon, something he assured her was inadmissible. Caroline tried to get some of Leibniz's writings translated into English. This effort brought her into contact with Samuel Clarke who had been recommended to her as a possible translator.

Clarke was a philosopher and theologian and also a disciple of Newton. In his *Being and Attributes of God,* dated 1704, Clarke had developed a proof of the existence of God. Caroline showed him a letter from Leibniz attacking certain of Newton's ideas and asked him to reply. This initiated a correspondence between the two men that continued until just a few days before Leibniz's death. Not surprisingly, there was no meeting of minds.

George Boole

From the point of view of our story, the most interesting fact about Samuel Clarke is that almost a century and a half after Leibniz's death, George Boole would demonstrate the efficacy of his own methods by using Clarke's proof of the existence of God as an example. In effect, with these methods, Boole succeeded in bringing to life part of Leibniz's dream. He had reduced Clarke's complicated deduction to a simple set of equations.[1]

In proceeding from the world of Leibniz and the seventeenth century European nobility to that of George Boole, we move forward not only two centuries in time, but also down several layers of social class. George, the first of four children, was born on November 2, 1815, in the town of Lincoln in the eastern part of England, to John and Mary Boole who had been childless for the first nine years of their marriage. John Boole, a cobbler who eked out only a meager living from his trade, had a passion for learning, and especially for scientific instruments. He proudly displayed a telescope he had made in his shop window. Unfortunately, he was not an effective business man and his talented, conscientious son soon found himself carrying the burden of supporting the whole family.[2]

In June 1830, the citizens of Lincoln were treated to a silly controversy in a local newspaper over the originality of an English translation of one of the poems of the ancient Greek writer Meleager. The translation had appeared in the *Lincoln Herald* as the work of "G. B. of Lincoln, aged 14 years," and one P. W. B. took the trouble to write accusing G. B. of plagiarism. P. W. B. admitted that he was unable to provide a reference to the source from which he was accusing G. B. of copying, but regarded it as simply beyond belief that the work could have been produced by a 14-year-old. The battle led to an exchange of several letters between G .B. and P. W. B., all duly published in the *Herald*.

George's family, who early recognized his ability, were far too poor to furnish him with a proper formal education, and so, with the help of his father, George was mainly self-taught. George studied not only Latin and Greek but also taught himself French and German and was able (much later, of course) to write mathematical research papers in these languages. George Boole never belonged to any particular religious denomination, and found it impossible to believe in the divinity of Christ, but throughout his life he held strong religious convictions. He soon abandoned his original ambition to join the clergy of the Church of England, in part because of his beliefs, but also because of his family's need for immediate financial help when his father's business collapsed. George was not yet 16 when he began his career as a teacher.

After two years at a small Methodist school some 40 miles from home he was fired, mainly it seems, owing to complaints about his irreligious behavior: he worked on mathematics on Sundays, and even in chapel! Indeed, it was at this time that Boole's efforts turned more and more to mathematics. In later years, reminiscing about this period in his life, he explained that having a very limited budget for buying books, he found that mathematics books provided the best value because it took longer to work through them than books on other subjects. He also liked to speak of the inspiration that suddenly came to him during his stay at the Methodist school. While walking across a field, the thought flashed across his mind that it should be possible to express logical relationships in algebraic form. This experience, which a biographer compares to that of Paul on the road to Damascus, was to bear fruit many years later.[3]

After teaching at the Methodist school, Boole took a position in Liverpool. But after six months of living and teaching there, he felt compelled to leave because of (in the words of his sister), "the spectacle of gross appetites and passions unrestrainedly indulged ..." presumably by the school headmaster.[4] His next job, in a village only four miles from home, was also of brief duration. This time, the reason was that, at the age of 19, concerned to put his family's finances on a sound basis, George Boole had decided to start his own school in his home town, Lincoln. For fifteen years, until accepting a professorship at a newly founded university at Cork, Ireland, Boole managed a successful career as a schoolmaster. His schools (there were three in succession) were the sole support of his parents and his siblings, although eventually his sister Mary Ann and brother William did participate in the work.

Although running a day and boarding school, and teaching numerous classes might be thought to be a full-time job, Boole managed during this period to make the transition from student of mathematics to creative mathematician. In addition, he somehow found time for activities of social improvement. He was a founder and trustee of a Female Penitent's Home in Lincoln whose purpose was "to provide a temporary home in which, by moral and religious instruction and the formation of industrious habits, females, who have deviated from the paths of virtue, may be restored to a reputable place in society." Boole's biographer speaks of prostitutes (who were evidently numerous in Victorian Lincoln) as the "penitent" women who were to be helped by this institution.[5] More likely, the typical client was a young woman of the servant class who found herself pregnant and abandoned after having been promised marriage by a lover of her own social class.* Some insight into George Boole's personal attitudes towards sexual matters may perhaps be gleaned from what he said in two of his lectures on non-mathematical subjects. In one, a lecture on education, he warned:

> A very large proportion of the extant literature of Greece and Rome ... is deeply stained with allusions and all too often with more than

*The study (Barret-Ducrocq, 1989) of a similar institution in London recounts many such tales of woe.

> allusions to the vices of Heathenism. ... But that the innocence of youth can be exposed to the contamination of evil without danger I do not believe.[6]

And a lecture on the proper uses of leisure (given after a successful campaign by the "Lincoln Early Closing Association" to obtain a ten-hour working day) included Boole's stern words:

> If you seek gratification in those pursuits from which virtue turns aside, you do so without excuse.[7]

Boole, following in his father's footsteps, was also deeply involved with the Lincoln Mechanics' Institute. These institutes, mainly devoted to after-hours education for artisans and other workers, had sprung up all over Victorian Britain. Boole did committee work for the one in Lincoln, made recommendations for improving the library, gave lectures, and provided teaching on a variety of subjects without remuneration.

Yet somehow, amidst all of this, he found time to study some of the most important English and continental mathematical treatises, and to begin making his own contributions. Much of Boole's early work bears witness to Leibniz's belief in the power of appropriate mathematical symbolism, of the manner in which the symbols seem to magically produce correct answers to problems almost unaided. Leibniz had pointed to the example of algebra. In England, as Boole began his own work, it was coming to be realized that the power of algebra comes from the fact that the symbols representing quantities and operations obeyed a small number of basic rules or laws. This implied that this same power could be applied to objects and operations of the most varied kind so long as they obeyed some of these same laws.[8]

In Boole's early work, he applied algebraic methods to the objects that mathematicians call *operators*. These "operate" on expressions of ordinary algebra to form new expressions. Boole was particularly interested in *differential operators*, so called because they involve the differentiation operation of the calculus mentioned in the previous chapter.[9] These operators were seen to be of particular importance because many fundamental laws of the physical universe take the form of differential equations, that is equations involving differential operators. Boole showed how certain differential equations could be solved by using methods of ordinary algebra applied to differential operators. Engineering and science students typically learn some of these methods in their sophomore or junior year in a course in differential equations.

During his years as a schoolmaster, Boole published a dozen research papers in the *Cambridge Mathematical Journal*. In addition, he submitted a very long paper to the *Philosophical Transactions of the Royal Society*. At first the Royal Society was loath to consider a submission from such an outsider, but finally decided to accept it, and later awarded it their Gold Medal.[10] Boole's method was to introduce a technique and then to apply it to a number of examples. He generally asked for no more in the way of *proof* that his methods were correct than that his examples worked out.[11]

At this time, Boole developed professional correspondences and friendships with a number of England's leading young mathematicians. A quarrel with the Scottish philosopher Sir William Hamilton that his friend Augustus De Morgan had fallen into brought Boole's thoughts back to his long ago flash of insight—that logical relationships might be expressible as a kind of algebra. Although Hamilton was an erudite scholar in aspects of metaphysics, he seems to have been something of a quarrelsome fool. Out of what can only have been his colossal ignorance of the subject, he published diatribes against mathematics. What had set him off was De Morgan's publication on logic that Hamilton claimed plagiarized what he thought of as his great discovery in logic, what he called the "quantification of the predicate." We need waste no time trying to understand this idea or the fierce controversy it generated—it is of importance only because of the stimulus it provided to George Boole.[12]

The classical logic of Aristotle that had so fascinated the young Leibniz involved sentences like:

1. All plants are alive.
2. No hippopotamus is intelligent.
3. Some people speak English.

Boole came to realize that what is significant in logical reasoning about such words as "alive," "hippopotamus," or "people" is the *class* or *collection* of all individuals described by the word in question: the *class* of living things, the *class* of hippopotamuses, the *class* of people. Moreover, he came to see how this kind of reasoning can be expressed in terms of an algebra of such classes. Boole used letters to represent classes just as letters had previously been used to represent numbers or operators. If the letters x and y stand for two particular classes, then Boole wrote xy for the class of things that are both in x and in y. As Boole himself put it:

> ... if an adjective, as "good," is employed as a term of description, let us represent by a letter, as y, all things to which the description "good" is applicable, i.e. "all good things," or the class "good things." Let it further be agreed, that by the combination xy shall be represented that class of things to which the names or descriptions represented by x and y are simultaneously applicable. Thus, if x alone stands for "white things," and y for "sheep," let xy stand for "white sheep;" and in like manner, if z stand for "horned things," ... let zxy represent "horned white sheep," ...[13]

Boole thought of this operation applied to classes like the operation of multiplication applied to numbers. However, he noticed a crucial difference: If once again y is the class of sheep, what is yy? It must be the class of things that are sheep and are also ... sheep. But this is the very same thing as the class of sheep; so $yy = y$. It is only a small exaggeration to say that Boole based his entire system of logic on the fact that when x stands for a class, the equation $xx = x$ is always true. We will return to this point later.*

George Boole was 32 when his first revolutionary monograph on logic as a form of mathematics was published. His more polished exposition, *The Laws of Thought* appeared seven years later. These were eventful years in Boole's life. Boole's social class and unconventional education had apparently ruled out his chances for an appointment at an English university. Strangely, it was the Irish "problem" that gave Boole an opening.

Among the many bitter complaints in Ireland concerning English rule was the Protestant character of its only university, Trinity College in Dublin. In response it was proposed by the British government to found three new universities to be called "Queen's Colleges" in Cork, Belfast, and Galway. Remarkably for the time, they would be non-denominational. Despite denunciations by Irish political and religious figures, who demanded institutions of a definitely Catholic character, the plans moved forward. Boole decided to apply for an appointment at one of these universities, and finally three years later, in 1849, he was appointed professor of mathematics at Queen's College in Cork.

By 1849, Ireland had come through the worst of the disaster of famine and disease brought by the potato blight, a devastating fungus that destroyed most of the potato crops on which the Irish poor depended. Many who did not starve to death were killed by the epidemics of typhus, dysentery, cholera, and relapsing fever to which their weakened immune systems had laid them open. The English rulers, slow to recognize the fungus as the underlying cause of the catastrophe, instead blamed the supposed indolence of the Irish. This social fiction was used to justify the continuing export of food from Ireland while millions went hungry and starved. Between 1845 and 1852, out of eight million Irish, at least a million died and another one and a half million emigrated.[14]

Boole had little to say about this: his strong expressions of indignation centered on cruelty to animals. Indeed, his attitude to the Irish people was rather equivocal as emerges from these lines from a sonnet to Ireland that Boole wrote just as the college in Cork was being inaugurated:

---

*Boole's equation $xx = x$ can be compared to Leibniz's $A \oplus A = A$. In both cases, an operation intended to be applied to pairs of items, when applied to an item and *itself*, yields that same item as a result.

> Yet thou in *wisdom* still art young, though old
> In misery and tears. Oh that thy store
> Of bitter thoughts, which brood upon the past,
> Were from thy bosom quite erased and worn.[15]

Although Cork was certainly no major intellectual or cultural center, the position provided Boole with the possibility of a life far more appropriate to his stature as one of the great mathematicians of the century than that of a schoolmaster. His father had recently died and, after making suitable provision for his mother, he was finally freed from the burden of being the family provider, and could think of having a personal life.

The mathematics taught at Cork was at a rather low level for a university. The syllabus began with "Fractional and Decimal Arithmetic" and continued with topics taught today in secondary school. Boole's annual salary was £250 in addition to a direct tuition fee of about £2 per term from each student. Since he had no assistant, he did all the grading of the weekly homework assignments.

Controversy over the Queen's Colleges continued. Although Cork's president was the distinguished Catholic scientist Sir Robert Kane, Catholics were certainly under-represented: of the academic staff of 21, only one other was Catholic. The Catholic Church hierarchy went so far as to forbid members of the clergy from participating in the work of the colleges. Some felt that Irish candidates for positions were deliberately passed over for relatively mediocre Englishmen or Scots. Nor did President Kane endear himself to his faculty. His wife had no wish to live in Cork, and so the President tried to run the college from Dublin. This, combined with his arbitrary pugnacious manner, led to one fight after another between the president and the faculty, sterile battles in which Boole usually found himself involved.[16]

Mary Everest, Boole's wife-to-be, later recounted some of her first impressions of the attitudes of some of the residents of Cork towards the man she would marry. One lady's answer to "What is the professor of mathematics like?" was "Oh he's like—the sort of man to trust your daughter with." Another lady explained the absence of her young children by informing Miss Everest that George Boole had taken them for a walk and that she was always happy when he walked with them. To the reply that Boole seemed to be a general favorite, the lady demurred:

> He is no favorite of mine, ... at least, I don't enjoy his society. I don't care to be with such very good people. ... he never shows you that he thinks you wicked, but when you are near anyone so pure and holy, you can't help feeling how shocked he must be at you. He makes me feel very wicked; but I am always at ease when the children are with him; I know they are getting some good.[17]

Mary Everest was the daughter of an eccentric clergyman and a niece of Lieutenant-Colonel Sir George Everest, whose name was given to the world's tallest mountain. She was also a niece of Boole's friend and colleague, John Ryall, Vice-President and Professor of Greek at Cork; this family connection brought George and Mary together. As a child Mary had displayed an aptitude for mathematics and after George began to tutor her, they grew to be good friends and frequent letter writers. It seems that Boole believed that their 17-year age difference precluded anything more, but five years after their first meeting when Boole was 40, matters came to a head with the death of Mary's father. As Mary was financially impoverished, George proposed at once, and they were married before the year was out.

Their marriage lasted a mere nine years, for Boole died at the age of only 49, after walking three miles to class in a cold October rainstorm. The ensuing bronchitis soon became pneumonia, and he died two weeks later. Tragically, his death may have been hastened by his wife's unorthodox medical views—apparently she treated his pneumonia by placing him between cold soaking bed sheets.[18]

The marriage had evidently been a very happy one.[19] Mary Boole recalled it "like the remembrance of a sunny dream." They had five children, all girls. Boole's widow lived well into the twentieth century, dying at the age of 84 while the First World War raged across the channel. She became attached to various systems of mystical belief and wrote a great deal of nonsense.

Boole's daughters all had interesting lives. The third daughter, Alicia, possessed a very remarkable geometric ability: she was able to visualize clearly geometric objects in four dimensions. This enabled her to make a number of important mathematical discoveries. However, the youngest daughter Ethel Lilian was the most astonishing. She was only six months old when her father died and she remembered her childhood as one of terrible poverty. Lily, as she was called, became involved with the Russian revolutionary emigres who made London their home during the late years of the nineteenth century. While on a trip to the Russian empire (which at that time included much of Poland) to help her revolutionary friends, she was seen by her future husband, Wilfred Voynich, from his prison cell, as she stared up at the Warsaw Citadel. Voynich recognized her years later after he had made his escape to London. This romantic beginning led to their marriage.

Lily became famous later as the author of *The Gadfly*, a novel inspired by her short but passionate love affair with Sidney Riley whose incredible life formed the basis for a television mini-series called *Riley: Ace of Spies*. With irony piled upon irony, Riley, a fervent anti-communist, was executed in Russia by the Bolsheviks, while his lover's novel, its true inspiration unknown, became required reading for Russian school children. In 1955 *Pravda* reported to its Moscow readers that the author of *The Gadfly* was alive and well in New York, and she received from Russia a royalty check for $15,000. She died five years later at the age of 96.[20]

George Boole's Algebra of Logic

Returning to Boole's new algebra applied to logic, we recall that if x and y represent two classes, Boole would write xy to stand for the class of those things that belong to both x and y. He intended the notation to suggest an analogy with multiplication in ordinary algebra. In contemporary terminology, xy is called the *intersection* of x and y.[21] We also saw that the equation $xx = x$ is always true when x represents a class. This led Boole to ask the question: *in ordinary algebra, where x stands for a number, when is the equation $xx = x$ true?* The answer is straightforward: the equation is true when x is 0 or 1 and for no other numbers. This led Boole to the principle that the algebra of logic was precisely what ordinary algebra would become if restricted to the two values 0 and 1. However, to make sense of this, it became necessary to reinterpret the symbols 0 and 1 as classes. A clue is provided by the behaviors of 0 and 1, respectively, with respect to ordinary multiplication: 0 *times any number is* 0; 1 *times any number is that very number.* In symbols,

$$0x = 0, \qquad 1x = x.$$

Now for classes, $0x$ will be identical to 0 for every x, if *we interpret* 0 *to be a class to which nothing belongs*; in modern terminology, 0 is the *empty set.* Likewise, $1x$ will be identical to x for every x, if 1 *contains every object under consideration,* or, as we may say, 1 is the "universe of discourse."

Ordinary algebra deals with addition and subtraction as well as multiplication. Thus, if Boole was to present the algebra of logic as just ordinary algebra with the special rule $xx = x$, he had to provide an interpretation for $+$ and $-$. So, if x and y represent two classes, Boole took $x + y$ to represent the class of all things to be found either in x or in y, nowadays called the *union* of x and y. Thus, to use Boole's own example, if x is the class of men and y is the class of women, then $x + y$ is the class consisting of all men and women. Also, Boole wrote $x - y$ for the class of things in x that are not in y.[22] If x represents the class of all people and y represents the class of all children, then $x - y$ would represent the class of adults. In particular, $1 - x$ would be the class of things not in x, so that

$$x + (1 - x) = 1.$$

Let us see how Boole's algebra works. Using ordinary algebraic notation, let us write x^2 for xx. So Boole's basic rule can be written as $x^2 = x$ or

$x - x^2 = 0$. Factoring this equation, following the usual rules of algebra,

$$x(1 - x) = 0.$$

In words: *nothing can both belong and fail to belong to a given class x.* For Boole, this was an exciting result, helping to convince him that he was on the right track. For as he said, quoting Aristotle's *Metaphysics*, this equation expresses precisely:

> ... that "principle of contradiction" which Aristotle has described as the fundamental axiom of all philosophy. "It is impossible that the same quality should both belong and not belong to the same thing ... This is the most certain of all principles ... Wherefore they who demonstrate refer to this as an ultimate opinion. For it is by nature the source of all the other axioms ... "[23]

Boole must have been delighted to obtain confirmation such as every scientist seeks when introducing new and general ideas: to see an important earlier landmark turn out to be a particular application of the new ideas, in this case Aristotle's principle of contradiction. In fact in Boole's time, it was common for writers on logic to equate the entire subject with what Aristotle had done so many centuries earlier. As Boole put it, "the science of Logic enjoys an immunity from those conditions of imperfection and of progress to which all other sciences are subject ..." The part of logic that Aristotle studied deals with inferences, called *syllogisms*, of a very special and restricted kind. They are inferences from a pair of propositions called *premises* to another proposition called the *conclusion*. The premises and conclusions must be representable by sentences of one of the following four types:*

Sentence type	**Example**
All X are Y.	All horses are animals.
No X are Y.	No trees are animals.
Some X are Y.	Some horses are pure-bred.
Some X are not Y.	Some horses are not pure-bred.

The following is an example of a valid syllogism:

$$\begin{array}{c} \text{All } X \text{ are } Y. \\ \underline{\text{All } Y \text{ are } Z.} \\ \text{All } X \text{ are } Z. \end{array}$$

*Lewis Carroll Carroll (1988, pp. 258–259) tells us that in a "sillygism" one proceeds from two "prim Misses" to a "delusion."

That this syllogism is *valid* means that whatever properties are substituted for X, Y, and Z, so long as the given two premises are true, the conclusion will be as well. Here are two instances of this syllogism:

All horses are mammals.	All boojums are snarks.
All mammals are vertebrates.	All snarks are purple.
All horses are vertebrates.	All boojums are purple.

Boole's algebraic methods can easily be used to demonstrate that this syllogism is valid. To say that everything in X also belongs to Y is the same as to say that there is nothing that belongs to X but not to Y, i.e., $X(1 - Y) = 0$ or equivalently $X = XY$. Likewise, the second premise can be written $Y = YZ$. Using these equations we get

$$X = XY = X(YZ) = (XY)Z = XZ,$$

the desired conclusion.[24]

Of course, not every proposed syllogism is valid. An example of an *invalid syllogism* can be obtained by interchanging the second premise with the conclusion in the previous example:

All X are Y.
All X are Z.
All Y are Z.

This time there is no way to use the premises $X = XY$ and $X = YZ$ to obtain the supposed conclusion $Y = YZ$.

In retrospect, it is difficult to understand the widespread belief that syllogistic reasoning constituted the whole of logic, and Boole was quite scathing in his denunciation of this idea. He pointed out that much ordinary reasoning involves what he calls *secondary propositions*, that is, propositions that express relations between other propositions. Such reasoning is not syllogistic.

For a simple example of such reasoning, let us listen in on a conversation between Joe and Susan. Joe can't find his checkbook and Susan is helping him.

SUSAN: Did you leave it in the supermarket when you were shopping?

JOE: No, I telephoned them, and they didn't find it. If I had left it there, they surely would have found it.

SUSAN: Wait a minute! You wrote a check at the restaurant last night and I saw you put your checkbook in your jacket pocket. If you haven't used it since, it must still be there.

JOE: You're right. I haven't used it. It's in my jacket pocket.

Joe looks and (if it's a good day for logic), the missing checkbook is there. Let us see how Boole's algebra could be used to analyze Joe and Susan's reasoning.

In their reasoning, Joe and Susan were dealing with the following propositions (each labeled with a letter):

L Joe left his checkbook at the supermarket.

F Joe's checkbook was found at the supermarket.

W Joe wrote a check at the restaurant last night.

P After writing the check last night, Joe put his check book in his jacket pocket.

H Joe hasn't used his check book since last night.

S Joe's checkbook is still in his jacket pocket.

They used the following pattern:

PREMISES. If L, then F
Not F
W & P
If W & P & H, then S
H

CONCLUSIONS. Not L
S

Like Aristotle's syllogisms, this pattern forms a valid inference. As with any valid inference, the truth of sentences called *conclusions* is inferred from the truth of other sentences called *premises*.

Boole saw that the same algebra that worked for classes would also work for inferences of this kind.[25] Boole used an equation like $X = 1$ to mean that the proposition X is true; likewise he used the equation $X = 0$

to mean that X is false. Thus, for "Not X," he could write the equation $X = 0$. Also, for X & Y he wrote the equation $XY = 1$. This works because X & Y is true precisely when X and Y are both true, while algebraically, $XY = 1$ if $X = Y = 1$, but $XY = 0$ if either $X = 0$ or $Y = 0$ (or both).

Finally, the statement "If X then Y" can be represented by the equation

$$X(1 - Y) = 0.$$

To see this, think of this statement as asserting

$$\textit{if } X = 1 \textit{ then } Y = 1.$$

But indeed, substituting $X = 1$ in the proposed equation leads to $1 - Y = 0$, that is, to $Y = 1$.

Using these ideas, Joe and Susan's premises can be expressed by the equations

$$\begin{aligned} L(1 - F) &= 0, \\ F &= 0, \\ WP &= 1, \\ WPH(1 - S) &= 0, \\ H &= 1. \end{aligned}$$

Substituting the second equation in the first, we get $L = 0$, the first desired conclusion. Substituting the third and fifth equations in the fourth, we get $1 - S = 0$, that is, $S = 1$, the other desired conclusion.

Now of course, Joe and Susan had no need for this algebra. But the fact that the kind of reasoning that ordinarily takes place informally and implicitly in ordinary human interactions could be captured by Boole's algebra encouraged the hope that more complicated reasoning could be captured as well. Mathematics may be thought of as systematically encapsulating highly complex logical inferences. This is part of the reason that mathematics is so useful in science. So an ultimate test of a theory of logic that aims at completeness is whether it encompasses all mathematical reasoning. We will return to this matter in the next chapter.

As a final example of Boole's methods, we turn to Samuel Clarke's proof of the existence of God mentioned at the beginning of this chapter. Without trying to follow Clarke's long complex deduction, it is at least amusing to see how Boole proceeds. We quote a small fragment:[26]

> The premises are:—
> 1st. Something is.

2nd. If something is, either something always was, or the things that now are have risen out of nothing.

3rd. If something is, either it exists in the necessity of its own nature, or it exists by the will of another being.

4th. If it exists in the necessity of its own nature, something always was.

5th. If it exists by the will of another being, then the hypothesis that the things which now are have risen out of nothing, is false.

We must now express symbolically the above propositions.
Let

x = Something is.
y = Something always was.
z = The things that now are have risen out of nothing.
p = It exists in the necessity of its own nature
(i.e., the *something* spoken of above).
q = It exists by the will of another being.

Boole then obtains from the premises the equations

$$\begin{aligned} 1 - x &= 0, \\ x\{yz + (1-y)(1-z)\} &= 0, \\ x\{pq + (1-p)(1-q)\} &= 0, \\ p(1-y) &= 0, \\ qz &= 0. \end{aligned}$$

One wonders what Clarke would have made of this reduction of his intricate metaphysical reasoning to manipulations of simple equations. Likely, as a disciple of Newton, he would have been pleased. On the other hand, the pugnacious metaphysician Sir William Hamilton who hated mathematics must have been horrified.

Boole and Leibniz's Dream

Boole's system of logic included Aristotle's and went far beyond it. But it still fell far short of what was needed to fulfill Leibniz's dream. Consider the following sentence:

All failing students are either stupid or lazy.

One might think of this sentence as

All X are Y.

However, this would require that the class of students being stupid or lazy be treated as a unit and would not permit any reasoning that sought to distinguish those who were failing because of stupidity from those who were failing because of laziness. In the next chapter we'll see how Gottlob Frege's system of logic does include reasoning of this subtler kind.

It is quite straightforward to use Boole's algebra as a system of rules for calculating, and say that, within its limits, it provided the *calculus ratiocinator* Leibniz had sought. Leibniz's writings on these matters were in the form of letters and other unpublished documents, and it was only late in the nineteenth century that a serious effort to gather and publish these was undertaken. So, there is no reasonable way that Boole could have been aware of his predecessor's efforts. Nevertheless it is interesting to compare Boole's full-blown system with Leibniz's fragmentary attempts.

Leibniz's fragment quoted in our first chapter included as its second axiom, $A \oplus A = A$. Thus the operation Leibniz considered was to obey Boole's fundamental rule: $xx = x$. Moreover, Leibniz proposed to present his logic as a full-fledged deductive system in which all of the rules are deduced from a small set of axioms. This is in accord with modern practice and shows Leibniz, in this respect, to have been ahead of Boole.

George Boole's great achievement was to demonstrate once and for all that logical deduction could be developed as a branch of mathematics. Although there had been some developments in logic after Aristotle's pioneering work (notably by the stoics in Hellenistic times and by the twelfth century scholastics in Europe), Boole had found the subject essentially as Aristotle left it two millennia earlier. After Boole, mathematical logic has been under continuous development to the present day.*

*An international organization, the *Association for Symbolic Logic* publishes two quarterly journals and holds regular meetings for the dissemination of new research. European logicians also hold annual meetings. New work on the relationships between logic and computers is presented at the annual international *Logic in Computer Science* and *Computer Science Logic* conferences.

CHAPTER 3

Frege: From Breakthrough to Despair

In June 1902 a letter arrived in Jena, a medieval German town, addressed to the 53-year-old Gottlob Frege from the young British philosopher Bertrand Russell. Although Frege believed that he had made important and fundamental discoveries, his work had been almost totally ignored. It must then have been with some pleasure that he read, "I find myself in agreement with you in all essentials ... I find in your work discussions, distinctions, and definitions that one seeks in vain in the work of other logicians." But, the letter continued, "There is just one point where I have encountered a difficulty." Frege soon realized that this one "difficulty" seemed to lead to the collapse of his life's work. It cannot have helped that Russell went on to write, "The exact treatment of logic in fundamental questions has remained very much behind; in your works I find the best I know of our time, and therefore I have permitted myself to express my deep respect to you."

Frege replied at once to Russell, acknowledging the problem. The second volume of his treatise in which he had applied his logical methods to the foundations of arithmetic was already at the printer, and he hastily added an appendix beginning with "There is nothing worse that can happen to a scientist than to have the foundation collapse just as the work is finished. I have been placed in this position by a letter from Mr. Bertrand Russell ..."

Many years later, more than four decades after Frege's death, Bertrand Russell had occasion to write:

> As I think about acts of integrity and grace, I realize that there is nothing in my knowledge to compare with Frege's dedication to truth. His entire life's work was on the verge of completion, much of his work had been ignored to the benefit of men infinitely less capable, his second volume was about to be published, and upon finding that his fundamental assumption was in error, he responded with intellectual pleasure clearly submerging any feelings of personal disappointment. It was almost superhuman and a telling indication of that of which men are capable if their dedication is to creative

> work and knowledge instead of cruder efforts to dominate and be known.[1]

Much of the contemporary philosopher Michael Dummett's work has been inspired by Frege's ideas. Yet when he wrote about Frege's integrity, it was in a quite different vein:

> There is some irony for me in the fact that the man about whose philosophical views I have devoted, over the years, a great deal of time to thinking, was, at least at the end of his life, a virulent racist, specifically an anti-semite. ... [His] diary shows Frege to have been a man of extreme right-wing opinions, bitterly opposed to the parliamentary system, democrats, liberals, Catholics, the French and, above all, Jews, who he thought ought to be deprived of political rights and, preferably, expelled from Germany. I was deeply shocked, because I had revered Frege as an absolutely rational man ...[2]

Frege's contributions were of immense importance. He provided the first fully developed system of logic that encompassed all of the deductive reasoning in ordinary mathematics, and his pioneering work using tools of logical analysis to study language provided the basis for major developments in philosophy. Today, under the subject heading "Frege, Gottlob" well over 50 items will be found in a typical university library. He died in 1925 a bitter man, believing that his life's work had led only to futility, his death ignored by the scholarly community.[3]

Gottlob Frege was born on November 8, 1848, in Wismar a small German town. His father, a theologian in the Evangelical faith, headed a girls' high school (where his mother was also employed). Frege was 38 when he married the 35-year-old Margarete Lieseberg who, after 17 years of marriage, died leaving no children behind.

At the request of a clergyman who was a relative on his mother's side, Frege adopted a five-year-old orphan in 1908. It was this son, Alfred, who brought to light the infamous diary Frege had kept in 1924, a year before his death, the diary that so outraged and disillusioned Michael Dummett. Alfred Frege as part of the German military occupation of Paris was killed in action in June 1944, a little over a week after the Allied landings in Normandy and just two months before the liberation of Paris. The diary had been typed by Alfred from his father's handwritten manuscript and in 1938, five years after Hitler had seized power, Alfred sent it to the Frege archive maintained by Heinrich Scholz. At that time the sentiments that so outraged Michael Dummett would have seemed unexceptional in Germany. The manuscript and a biography Alfred had written of his father are lost.

Frege was 21 when he entered the university. After two years at Jena he moved to Göttingen University where, three years later, he received a Ph.D. in mathematics. Then, he obtained an appointment as lecturer

GOTTLOB FREGE
(Institute for Mathematical Logic and Foundational Research, Münster University)

("Privatdozent") at the University of Jena, a position without salary. It seems that Frege was supported at this time by his mother who, on his father's death, had taken over management of the girls' school. After five years Frege was appointed Associate Professor ("Ausserordentlichen Professor") at Jena where he remained until his retirement in 1918. Because his colleagues didn't really value his work, he was never promoted to a full professorship. His death at Bad Kleinen near Wismar, where his impoverishment had forced him to board with relatives, came little over a year after the final entry in his deplorable diary.

In 1873, the year of Frege's initial appointment at Jena, Germany, newly united, was in a state of euphoria. The war against the France of Napoleon III had ended in a great victory. Industry was expanding at breakneck speed. Until the death of Kaiser Wilhelm I, his Chancellor, Bismarck, continued his cunning policy of maintaining the security of Germany by means of a carefully nurtured system of alliances. Bismarck and the "old Kaiser" remained heroes to Frege for his entire life. However, Bismarck was a reactionary who saw to it that the emperor maintained total control of military affairs and foreign relations. He regarded democracy as anathema, and pushed legislation outlawing many of the activities of the Social Democratic party.

Soon after Wilhelm II succeeded to the throne, he got rid of Bismarck. The new Kaiser, a vainglorious and insecure man, oversaw a disastrous foreign policy. Repeatedly misjudging the effect of his maneuvers, he managed to so alarm the other European powers that France, Russia, and England formed an alliance against Germany. Faced with the danger of a war on two fronts, against Russia on the east and against France on the west, the German general staff produced the clever, but ultimately disastrous, Schlieffen plan, designed to defeat France quickly before Russia could complete its ponderous mobilization.[4]

So when, with German encouragement, the Austrians attacked Serbia in the summer of 1914, in response to the assassination of Archduke Ferdinand, and Russia began mobilization to stress its determination that Austria not be permitted to destroy fellow Slavs, the German generals explained to the Kaiser that the Schlieffen plan calling for a German attack through Belgium had to be implemented at once. The attendant violation of Belgium's neutrality brought England into this catastrophic war whose consequences cast their shadow on the entire twentieth century. In war things rarely go according to plan, and when the Schlieffen plan attack petered out, the fighting degenerated into a murderous stalemate, slaughtering the best part of a generation of European men in trench warfare. Seemingly unaware that the fighting was going badly, many German academics called for a peace in which Germany would annex much territory, including all of Belgium.

As victory continued to elude the Germans and the English siege took its toll, the military command was put into the hands of General Ludendorff. This capricious gambler (who was later to participate in Hitler's beer hall "Putsch") refused to consider a compromise peace until a British break-through in the Balkans threatened to roll up the German flank. With defeat staring him in the face, Ludendorff told the Kaiser that an armistice was essential. So ended the war and the German monarchy.

The government that assumed power in the new German republic was Social Democratic, and many Germans (Frege among them) came to accept the story that Germany had been forced into the war against its will, had not been defeated, but had been betrayed by the socialists, and (many were soon adding) the Jews. This was the poisonous atmosphere that ultimately made it possible for Hitler to assume power.

The year 1923 saw the great post-war hyper-inflation in Germany, in part the result of the unrealistic reparations imposed by the Versailles treaty. This financial catastrophe wiped out the values of personal savings, and presumably, of Frege's pension. It was in this situation that Frege produced his terrible diary. He looked for a great leader to rescue Germany from the lowly position into which it had been thrust. Having held high hopes for Ludendorff to play this role, he was disappointed that he had joined Hitler's Putsch. He still had hope that General Hindenburg might be the leader, but feared that he was too old; Frege did not live to see Hindenburg hand the keys to the republic to Adolf Hitler.

In his diary entry for April 22, 1924, Frege reminisces about a time when the Jews of his home town were treated in what he thought was an appropriate manner and also manages to disclose his views on the French and their baleful influence:

> There was a law at that time that Jews were permitted to stay overnight in Wismar only in the time of certain annual fairs, ... I suppose this decree was old. The old Wismarkers must have had experiences with the Jews that had led them to this legislation.
>
> It must have been the Jewish way of doing business together with the Jewish national characteristics that is tied together closely with the way of doing business. ... There came universal suffrage, even for Jews. There came the freedom of movement, even for Jews, presents from France. We make it so easy for the French to bless us with gifts. If one had only turned to noble and patriotic Germans ... The French had treated us nastily enough indeed before 1813, and nevertheless we have this blind admiration of all things French. ... I have only in the last years really learned to comprehend antisemitism. If one wants to make laws against the Jews, one must be able to specify a distinguishing mark by which one can recognize a Jew for certain. I have always seen this as a problem.

The problem, merely theoretical for Frege, of defining Jews with sufficient precision so that one could make laws against them, became quite a practical problem under the Nazis. Ludwig Wittgenstein, thought to be one of the great thinkers of the twentieth century and an admirer and disciple of Frege, would have qualified as a Jew under the Nazi racial code.

Other diary entries rail against the Social Democrats and Catholics:

> The Reich suffered from a cancer in 1914, namely Social Democracy. (April 24)
>
> To be sure, I regarded Ultramontanism and its embodiment in the Zentrum as very detrimental for our Reich and nation; nevertheless, the revelations of ... Ludendorf in his [recent] article on the efforts and machinations of the ultramontanes give me insights which have most deeply disturbed me.* I implore anybody who does not yet believe in the thoroughly unGerman spirit of the Zentrum to read and reflect on the stated article of His Excellency Ludendorf ... This is the most evil enemy which undermined Bismarck's Reich. ... [The Ultramontanes] will always look to the Pope to get their instructions. (April 26)[5]

Frege's extreme right-wing ideas were hardly rare in Germany after World War I. Nevertheless, we may wonder whether the diary represents only the thoughts of a bitter (and possibly senile) old man within a year of his death. Alas, there is little doubt that Frege had held right-wing views for some time. Frege's colleague, Bruno Bauch, a philosophy professor at Jena, founded a right-wing philosophical society (the DLG) during the war, and he edited its journal. Frege was one of the early adherents of the DLG, and published in its journal. Bauch's writings on the concept of nation insisted that no Jew could really be a German. His group came out in full support of the Nazis when they took power in 1933.[6]

Frege's *Begriffsschrift*

It is with a sense of relief that one turns from Frege's awful views, expressed as his life drew towards its end, to the brilliant contributions he made as a young man. In 1879,* he published a booklet of fewer than 100 pages entitled *Begriffsschrift,* a hard-to-translate word Frege constructed from the German words *Begriff* ("concept") and *Schrift* (roughly "script" or "mode of writing"). It was subtitled, "a formula language, modeled upon

*The Zentrum party was oriented towards the Catholic Church. Its "Ultramontanism" referred to the influence from "over the mountains," that is, Rome.

*I was invited to present an address at a scientific conference in 1979 commemorating the hundredth anniversary of the *Begriffsschrift* in which I was to trace its consequences for computer science. This was the beginning of my second career as a historian of science.

that of arithmetic, for pure thought." This work has been called "perhaps the most important single work ever written in logic."[7]

Frege sought a system of logic that included all of the deductive inferences in mathematical practice. Boole took ordinary algebra as his starting point and used the symbols of algebra to represent logical relations. Since Frege intended algebra, like other parts of mathematics, to be built as a superstructure with his logic as a foundation, he regarded it as important to introduce his own special symbols for logical relationships to avoid confusion.

Also, where Boole had thought of propositions that express relations between other propositions as "secondary propositions," Frege saw that the same relations that connect propositions can also be used to analyze the structure of individual propositions, and he made these relations the basis of his logic. This crucial insight has gained general acceptance and forms the basis of modern logic.

For example, Frege would analyze the statement that "All horses are mammals" using the logical relationship *if* ... *then* ...:

if x is a horse, *then* x is a mammal.

Likewise, he would analyze the statement that "Some horses are pure-bred" using the logical relationship ... *and* ...:

x is a horse *and* x is pure-bred.

However, the letter x is used differently in these two examples. In the first example one wants to say that what is asserted is true *whatever* x *might be,* that is, *for every* x. But in the second example what is wanted is only the assertion *for some* x. In the symbolism in current use, *for every* is written $\forall$ and *for some* is written $\exists$. So, the two sentences could be written as follows:

$$(\forall x)(\textit{if } x \text{ is a horse, } \textit{then } x \text{ is a mammal})$$
$$(\exists x)(x \text{ is a horse } \textit{and } x \text{ is pure-bred})$$

The symbol $\forall$, an upside-down A, suggests "all" and is called a *universal quantifier.* Likewise the symbol $\exists$, a backwards E, is called an *existential quantifier*, and is intended to suggest "exists." So this second sentence could be read:

There exists x such that x is a horse *and* x is pure-bred.

The logical relation *if* ... *then* ... is usually symbolized $\supset$, and the relation ... *and* ... is symbolized $\wedge$. Using these symbols, the sentences become:[8]

$$(\forall x)(\ x \text{ is a horse } \supset\ x \text{ is a mammal})$$
$$(\exists x)(\ x \text{ is a horse } \wedge\ x \text{ is pure-bred})$$

They can be abbreviated as follows:

$$(\forall x)(\text{ horse}(x) \supset \text{ mammal}(x))$$
$$(\exists x)(\text{ horse}(x) \wedge \text{ pure-bred}(x))$$

Or more briefly:

$$(\forall x)(h(x) \supset m(x))$$
$$(\exists x)(h(x) \wedge p(x))$$

Joe and Susan's effort to use logic in locating Joe's wallet was used as an example in the previous chapter. In that example, we used letters to abbreviate sentences as follows:

L Joe left his checkbook at the supermarket.

F Joe's checkbook was found at the supermarket.

W Joe wrote a check at the restaurant last night.

P After writing the check last night, Joe put his check book in his jacket pocket.

H Joe hasn't used his check book since last night.

S Joe's checkbook is still in his jacket pocket.

Their reasoning came down to the following pattern:

PREMISES. If L, then F
Not F
W & P
If W & P & H, then S
H

CONCLUSIONS. Not L
S

Using the symbol $\neg$ to stand for "not," and the other symbols we've introduced, this now becomes

$$\begin{array}{c} L \supset F \\ \neg F \\ W \wedge P \\ W \wedge P \wedge H \supset S \\ H \\ \hline \neg L \\ S \end{array}$$

One final symbol should be mentioned: $\vee$ standing for ... *or* .... The following table provides a summary of the symbols that have been introduced:

$\neg$	not ...
$\vee$	... or ...
$\wedge$	... and ...
$\supset$	if ... then ...
$\forall$	every
$\exists$	some

At the end of the previous chapter, the statement that *all failing students are either stupid or lazy* was exhibited as an example whose logical structure would be missed by Boole's analysis. In Frege's logic, it is easy. Writing

$$\begin{array}{lll} F(x) & \text{for} & x \text{ is a failing student,} \\ S(x) & \text{for} & x \text{ is stupid,} \\ L(x) & \text{for} & x \text{ is lazy,} \end{array}$$

the sentence can be expressed as

$$(\forall x)(F(x) \supset S(x) \vee L(x)).$$

By now it should be clear that Frege was not just developing a mathematical treatment of logic, but was creating a new language. In this he was guided by Leibniz's notion of a universal language that would gain its power from a judicious choice of symbols.[9] The expressiveness of this language can be gauged from the following examples using $L(x, y)$ to stand for x *loves* y.

Everyone loves someone.	$(\forall x)(\exists y)\, x$ loves y	$(\forall x)(\exists y)L(x,y)$
Someone loves everyone.	$(\exists x)(\forall y)\, x$ loves y	$(\exists x)(\forall y)L(x,y)$
Everyone is loved by someone.	$(\forall y)(\exists x)\, x$ loves y	$(\forall y)(\exists x)L(x,y)$
Someone is loved by everyone.	$(\exists y)(\forall x)\, x$ loves y	$(\exists y)(\forall x)L(x,y)$

Here is one more example:

Everyone loves a lover.

As a first stab we write:

$$(\forall x)(\forall y)[y \text{ is a lover } \supset L(x,y)].$$

Now, if we construe being a lover as simply meaning loving someone, we can replace *y is a lover* by $(\exists z)L(y,z)$, finally obtaining:

$$(\forall x)(\forall y)[(\exists z)L(y,z) \supset L(x,y)].$$

Frege Invents Formal Syntax

Boole's logic was simply another branch of mathematics to be developed using ordinary mathematical methods. This of course includes using logical reasoning. But there is something circular about *using* logic to *develop* logic. For Frege this was unacceptable. His intention was to show how all mathematics could be based on logic; logic was to provide a *foundation* for all the rest of mathematics. For this to be at all convincing, Frege had to find some way to develop his logic without *using* logic in the process.

His solution was to develop his *Begriffsschrift* as an artificial language with mercilessly precise rules of grammar, or as one says, of *syntax*. This made it possible to exhibit logical inferences as purely mechanical operations, so-called *rules of inference*, having reference only to the patterns in which symbols are arranged. It was also the first example of a formal artificial language constructed with a precise syntax. From this point of view, the *Begriffsschrift* was the ancestor of all computer programming languages in common use today.

The most fundamental of Frege's rules of inference works like this: if $\Diamond$ and $\triangle$ are any two sentences of Frege's *Begriffsschrift*, then if $\Diamond$ and $(\Diamond \supset \triangle)$ are both asserted, then one is permitted to also assert the sentence $\triangle$. It is important to notice that to carry out this operation, no understanding of what $\supset$ means is required. Of course we can see that the rule cannot lead to error because it only enables one to proceed from $\Diamond$ and (*if* $\Diamond$ *then* $\triangle$) to $\triangle$. But to actually employ the rule, it is only necessary to match up the individual symbols constituting the sentence $\Diamond$ with symbols in the first part of the longer sentence.[10] In our example of locating Joe's wallet, we had the premise

$$W \wedge P \wedge H \supset S.$$

If we were able to also assert $W \wedge P \wedge H$, then the rule would enable us to also assert one of the desired conclusions, namely S. Here is how the match-up would go:

$$W \wedge P \wedge H \supset S.$$
$$W \wedge P \wedge H.$$

Frege's logic has become the standard taught to undergraduate students in logic courses in mathematics, computer science, and philosophy departments.[11] It has been the basis for an enormous body of research, and indirectly led Alan Turing to formulate the idea of an all-purpose computer. But this is getting ahead of ourselves.

Frege's logic was an enormous advance over Boole's. For the first time an exact system of mathematical logic encompassed, at least in principle, all the reasoning ordinarily used by mathematicians. But in attaining this goal, something was given up. Beginning with some premises in Frege's logic, Frege's rules could be applied in an attempt to reach a desired conclusion. But if the attempt failed, Frege provided no means to know whether this was because not enough cleverness or persistence was employed, or whether the desired conclusion simply did not follow from the given premises. This lack meant that Frege's logic did not fulfill Leibniz's dream that with the words "Let us calculate," those knowing the rules of logic would be able to proceed to determine unfailingly whether or not some conclusion follows.

Why Bertrand Russell's Letter Was So Devastating

If Frege's logic was such a great achievement, why did Russell's letter lead Frege to despair? Frege regarded his logic as only a stepping stone towards complete foundation for arithmetic. Although the differential and integral calculus of Leibniz and Newton led to fruitful developments, there were serious problems in justifying some of the steps in the reasoning mathematicians were in the habit of employing. During the nineteenth century these problems were gradually cleared up, ultimately by developing a new and profound theory of the number system of mathematics. However, in the end we still rely on the so-called *natural* (or *counting*) numbers:

$$1, 2, 3, \ldots$$

Frege wanted to provide a purely logical theory of the natural numbers and thereby to demonstrate that arithmetic, and indeed all of mathematics including developments stemming from the differential and integral calculus, could be regarded as a branch of logic. This point of view, which came to be called *logicism*, was also that of Bertrand Russell. Logicism has been explained by the American logician Alonzo Church as maintaining that the relationship between logic and mathematics is that between the elementary and the advanced part of one and the same subject.*

Thus Frege wanted to be able to define the natural numbers in purely logical terms, and then to use his logic to derive their properties. The number 3 for example was to be explained as part of logic. How could this be possible? A natural number is a property of a set, namely, the number of its elements. The number 3 is something that all of the following have in common: the Holy Trinity, the set of horses pulling a troika, the set of leaves on a (normal) clover leaf, the set of letters $\{a, b, c\}$. Without saying anything about the number 3, one can see that any two of these sets have the *same* number of elements. We can simply match them up. Frege's idea was to identify the number 3 with the collection of all of these sets. That is, the number 3 is just the set of all triples. In general, the number of elements in a given set can be *defined* to be the collection of all those sets that can be matched one-to-one with the given set.[12]

Frege's two-volume treatise on the foundations of arithmetic showed how to develop the arithmetic of natural numbers using the logic developed in his *Begriffsschrift*. Bertrand Russell's letter of 1902 showed Frege that this entire development was inconsistent, that is, self-contradictory. Frege's arithmetic, in effect, made use of sets of sets. Russell showed in his letter that reasoning with sets of sets can easily lead to contradictions.

Russell's "paradox" can be explained as follows: Call a set *extraordinary* if it is a member of itself; otherwise call it *ordinary*. How could a set be extraordinary? Russell's own example of an extraordinary set is: *the set of all those things that can be defined in fewer than 19 English words.* Since we have just defined this set using only 16 words, it belongs to itself and therefore is extraordinary. Another example is the set of all things that are not sparrows. Whatever this set might be, it is surely not a sparrow. So this set too is extraordinary.

Russell proposed to Frege the set $\mathcal{E}$ of all ordinary sets. Is $\mathcal{E}$ ordinary or extraordinary? It must be one or the other. But it seems to be neither. Could $\mathcal{E}$ be ordinary. If so, since $\mathcal{E}$ is the class of all ordinary sets, it would belong to itself. But then it would be extraordinary. OK. Then $\mathcal{E}$ would have to be extraordinary. Therefore, it would not belong to itself, since it is the set of ordinary sets. But that would make it ordinary! Either way one is led to a contradiction!

*It is now generally recognized that, by the use of numerical coordinates, geometry can also be reduced to arithmetic. However, Frege always believed that geometry had to be regarded as separate. I'm indebted to Patricia Blanchette for emphasizing this aspect of Frege's thought and for other helpful comments on this section.

Russell's paradox is first cousin to a large number of puzzles that are simply amusing. But when Frege received Russell's letter, he was not amused. He realized at once that the contradiction could be readily derived within the system he was using to develop arithmetic. Now, a mathematical proof that runs into a contradiction is a demonstration that one of the premises of the argument was false. This principle is used all the time as a useful proof method: to prove a proposition, one shows that its denial leads to a contradiction. But for poor Frege, the contradiction had shown that the very premises on which his system was built were untenable. Frege never recovered from this blow.[13]

Frege and the Philosophy of Language

In 1892 Frege published a paper in a philosophical journal whose title may be translated as *On Sense and Denotation.*[14] Along with Frege's logic, it is because of the issues raised in this paper that philosophers have been so interested in his work.

Frege pointed out that different words may be used to denote one specific object although they may have quite different senses or meanings. His famous example uses the phrases "the morning star" and "the evening star." Their *sense* is quite different: one is the bright star seen after sunset, the other the one seen before sunrise. But both *denote* the same planet, Venus. The fact that both phrases refer to the same object is not obvious; it was at one time a real astronomical discovery. Some of Frege's concerns have to do with substitutivity: Consider the sentence

Venus is the morning star.

This is very different from

Venus is Venus.

This is the case although in fact, one sentence was derived from the other by replacing one phrase by another denoting the same object.

These ideas represent the beginning of a major branch of twentieth century philosophy: the philosophy of language.[15] In addition, some key concepts in contemporary computer science may be said to have their origin in this same essay.[16]

Frege and Leibniz's Dream

Frege thought of his *Begriffsschrift* as embodying the universal language of logic that Leibniz had called for. Indeed, Frege's logic can deal with the most diverse subjects. But to Leibniz it would likely have been a disappointment. It fell short of his desires in at least two important respects.

Leibniz had imagined a language that was capable not only of logical deduction but that also would automatically include all the truths of science and of philosophy. This naive expectation was only conceivable before the massive development of science in the eighteenth and nineteenth centuries based on careful experiment as well as theorizing.

From the point of view of our story, it is more appropriate to point to a different limitation of Frege's logic. Leibniz had called for a language that would also be an efficient instrument of calculation, one that would enable logical inferences to be carried out systematically by the direct manipulation of symbols. In fact any but the simplest of deductions are almost unbearably complicated in Frege's logic. Not only are such deductions tediously long, but Frege's rules provide no calculational procedures for determining whether some desired conclusion can be deduced from given premises in the logic of his *Begriffsschrift*.

Because the *Begriffsschrift* did fully encapsulate the logic used in ordinary mathematics, it became possible for mathematical activity to be investigated by mathematical methods. As we will see, these investigations led to some very remarkable and unexpected developments. The search for a calculational method that could show whether or not a proposed inference in Frege's logic is correct reached its climax in 1936 with a *proof* that no such general method exists.

This was bad news for Leibniz's dream. However, in the process of proving this negative result, Alan Turing discovered something that would have delighted Leibniz: he found that it was possible, in principle, to devise one single "universal" machine that could alone carry out any possible computation.

CHAPTER 4

Cantor: Detour through Infinity

The sequence of numbers $1, 2, 3, \ldots$, the so-called *natural* or *counting* numbers, goes on forever. No matter how large a number you start with, you can always get a larger number by adding 1. One may conceive of the natural numbers as generated by a process, beginning with 1 and successively adding 1:

$$1 + 1 = 2, \;\; 1 + 2 = 3, \;\; \ldots, \;\; 1 + 99 = 100, \;\; \ldots$$

Such a process, continuing beyond any finite bound, was characterized by Aristotle as a "potential infinity." However, Aristotle was not willing to accept as legitimate the culmination of this process: the infinite set of all natural numbers. This would be a "completed" or "actual" infinity, and Aristotle declared that such were illegitimate.[1] Aristotle's views heavily influenced the scholastic religious philosophers of the twelfth century, particularly Thomas Aquinas. The problem of the nature of the infinite has been perplexing for mathematicians, philosophers, and theologians alike. Theologians could propose that a "completed" infinity was actually an aspect of God, and conclude that for mere humans it had to remain a mystery. Leibniz was not put off by such considerations, writing:

> I am so in favor of the actual infinite that instead of admitting that Nature abhors it, as is commonly said, I hold that Nature makes frequent use of it everywhere, in order to show more effectively the perfections of its Author.[2]

The limit processes of the calculus that became so important for mathematics in the eighteenth and nineteenth centuries exemplified potential infinity. In this connection, the great German mathematician Carl Friedrich Gauss (1777–1855) warned:

> ...I protest above all against the use of an infinite quantity as a *completed* one, which in mathematics is never allowed. The infinite is only a manner of speaking, in which one properly speaks of limits.[3]

After the middle of the nineteenth century mathematical problems that arose quite naturally out of current concerns seemed to call for the use of completed infinities in their precise formulation. Among the mathematicians who were coping with this situation, it was only Georg Cantor who, flying in the face of Gauss's warning, accepted the challenge to create a profound and coherent mathematical theory of the actual infinite. Cantor's work unleashed a storm of criticism: not only mathematicians, but also philosophers and theologians attacked the temerity of one who would bring the methods of mathematical science to bear on the hitherto sacrosanct domain of the infinite. Frege was supportive of Cantor's embrace of the actual infinite, recognizing its importance for the future of mathematics. Frege also saw quite clearly that a stormy struggle would develop between those mathematicians who embraced Cantor's infinite and those who regarded it as anathema:

> For the infinite will eventually refuse to be excluded from arithmetic .... Thus we can foresee that this issue will provide the setting for a momentous and decisive battle.[4]

What Frege could not have foreseen as he wrote these lines was that the very foundation for arithmetic that he himself had developed would be an early casualty of that battle, a victim of the paradox that Bertrand Russell would call to his attention a decade later in that famous letter, a paradox that Russell would find while exploring the implications of Cantor's infinite. And Frege could certainly never have imagined that the ensuing tumultuous discussions, investigations, and disputes over Cantor's infinite would one day provide key insights leading to the development of all-purpose digital computers.

Engineer or Mathematician

In an unlikely setting for a future professor of mathematics at a German university, Georg Cantor was born in 1845 in St. Petersburg, Russia. Cantor's mother, Marie Böhm came from a distinguished musical family, and she herself was an accomplished musician. His father, Georg Waldemar Cantor, was born in Copenhagen, but was brought to St. Petersburg as a child. It is believed that he was raised and educated there in a Lutheran Evangelical mission. Although Marie had been baptized a Roman Catholic, she also adhered to the Evangelical Church after her marriage, and Georg Cantor and his three siblings were raised in that faith.[5]

GEORG CANTOR
(Bildarchiv des Mathematischen Forschungsinstituts Oberwolfach)

Georg Waldemar Cantor was a very successful business man. He worked as a wholesaling agent in St. Petersburg, and later became a broker at the St. Petersburg Stock Exchange. One author, referring to the letters Georg had received from his father while a student, was moved to write:

> ...one is fascinated by this multifaceted, cultivated, mature, and kind individual. They [the letters] breathe a spirit not always found among successful business men.[6]

Although tuberculosis, the nineteenth century's great plague, hit poor neighborhoods with particular force, the rich were not immune. Georg's father contracted this dread disease and ultimately died of it. Although still in his forties, his illness led Georg Waldemar to liquidate his business and move his family to Germany when his son was 11. But his success had been such that, even after his death seven years after the move, his four children were very well provided for.

Georg's father had believed that engineering was the profession most appropriate to his son's talents, but, to Georg's great joy, he finally acquiesced in the boy's desire to be a mathematician. In Berlin, the young Georg Cantor had the opportunity to study under three great mathematicians: Karl Weierstrass, Ernst Kummer, and Leopold Kronecker. Cantor's mathematical interests began in quite traditional areas. It would have been difficult to predict at the beginning of his career that he was destined to expand the horizons of mathematical thought in a revolutionary direction. Nor that his teacher, Kronecker, would refuse to accept his new mathematics of the infinite as legitimate.

Halle, where Cantor assumed his first university position and where he was to spend the rest of his life, was an industrial city 35 miles up the Saale River from Frege's home in Jena. Quite typical for a beginning academic career in Germany at that time, Cantor was appointed a Privatdozent, a lecturer without pay. Obviously, under these circumstances, independent financial resources were necessary for launching an academic career. The leading mathematician at Halle, Eduard Heine, recognized Cantor's great mathematical powers, and persuaded him to work on some problems involving infinite series. In the first chapter, we have already encountered Leibniz's famous infinite series:

$$\frac{\pi}{4} = 1 - \frac{1}{3} + \frac{1}{5} - \frac{1}{7} + \frac{1}{9} - \frac{1}{11} + \cdots .$$

The "infinities" encountered in such *series* are potential infinities only, exactly the sort Gauss (quoted above) had in mind: " [infinite] only in a manner of speaking, in which one properly speaks of limits." For an infinite series, one seeks a *limit* to which one gets ever closer as one adds more and more terms (in the case of Leibniz's series, this limit is $\frac{\pi}{4}$); one says that the series *converges* to the limit. There is no question of a "completed" infinity; at any stage in the process one has simply added finitely many numbers.

Naturally, the subject of infinite series had advanced considerably in the two centuries since Leibniz's time. Cantor studied *trigonometric series*,[7] (so-called because the terms involve the *sine* and *cosine* from trigonometry). He wanted to find out under what circumstances two different series of this type could converge to the same thing, and in fact, to prove that such circumstances would be very unusual. This investigation took Cantor far afield: he found that in order to get the desired results, he had to treat infinite sets as "completed" wholes and to perform complex operations on them. Soon, he was developing the theory of sets (*Mengenlehre* in German) as an autonomous subject.

Infinite Sets Come in Different Sizes

Granting that it makes sense to deal with the set of all natural numbers, $1, 2, 3, \ldots$, as an example of a "completed, actual" infinity, does it also make sense to ask: how many numbers are there in this set? Are there infinite numbers that can be used to "count" infinite sets? Leibniz, who had no objection to completed infinities as such, considered this question in a letter to the Catholic priest, theologian, and philosopher Nicolas Malebranche. His conclusion was that such infinite numbers do not exist.

We may explain his reasoning as follows: we can tell that two sets have the same number of members, without even knowing what that number is, by matching the elements of one of the sets in a one-one manner with those of the other set.* For example, if one observes that there are no empty seats and no standees in an auditorium, then one can conclude (without counting) that the number of people in the audience and the number of seats are the same—one is matching up each seat with the person occupying it. Leibniz held that if such things as infinite numbers did exist, then the same idea should apply to them: if a one-one matching can be defined between two infinite sets, then one should be able to conclude that the two sets have the same number of members.

Then, he proposed to apply this concept to the following two sets: the set of all natural numbers $1, 2, 3, \ldots$ and the set of even natural numbers $2, 4, 6, \ldots$. It is easy to devise a one-to-one matching between these two sets by simply matching, with each natural number its double, like this:

$$\begin{array}{ccccc} 1 & 2 & 3 & 4 & \ldots \\ \updownarrow & \updownarrow & \updownarrow & \updownarrow & \\ 2 & 4 & 6 & 8 & \ldots \end{array}$$

*This is the idea Frege invoked in his thwarted attempt to *define* "number."

Notice that even though the sets are infinite, the specified matching between the set of all natural numbers and the set of even numbers is perfectly explicit. For example, corresponding to the natural number 117 is the even number 234; corresponding to the natural number 4228 is the even number 8456, etc. Leibniz reasoned that if there were such things as infinite numbers, the existence of this match-up would force us to conclude that the "number" of natural numbers is the same as the "number" of even numbers. But how could this be?

Among the natural numbers, are not only the even numbers themselves, but also all of the odd numbers, themselves constituting an infinite set. And one of the most basic mathematical principles, going back to Euclid, is that the "whole" is greater than any of its parts.[8] Hence Leibniz concluded that the very concept of the "number" of all natural numbers is incoherent, that it makes no sense to speak of the number of elements in an infinite set. As he put it:

> For any number there exists a corresponding even number which is its double. Hence the number of all numbers is not greater than the number of even numbers, that is, the whole is not greater than the part.[9]

Cantor reasoned much as Leibniz had and faced the same dilemma: *either* it makes no sense to speak of the *number* of elements in an infinite set *or* some infinite sets will have the same number of elements as one of their subsets. However, while Leibniz had chosen one horn of this dilemma, Cantor chose the other. He went on to develop a theory of *number* that would apply to infinite sets, and just accepted the consequence that an infinite set could have the same number of elements as one of its parts.

Starting where Leibniz had left off, Cantor began studying when it was possible to set up one-to-one matchings between two different infinite sets. Leibniz had found that a one-to-one matching could be established between the set of natural numbers and one of its subsets, the even numbers. Cantor considered sets that seemed to be larger than the set of natural numbers.

One example he thought about was the set of numbers that can be represented as (positive) fractions,[10] like $\frac{1}{2}$ or $\frac{5}{3}$. Since natural numbers could be represented by fractions with the denominator 1 (like $\frac{7}{1}$), the set of natural numbers can be regarded as a subset of this set. But, with a little thought, Cantor found that he could set up a one-to-one matching between the set of these fractions and the set of natural numbers. The fractions can be arranged in a sequence like this:

$$\left|\ \frac{1}{1}\ \right|\ \frac{1}{2}\ \frac{2}{1}\ \left|\ \frac{1}{3}\ \frac{2}{2}\ \frac{3}{1}\ \right|\ \frac{1}{4}\ \frac{2}{3}\ \frac{3}{2}\ \frac{4}{1}\ \left|\ \frac{1}{5}\ \frac{2}{4}\ \frac{3}{3}\ \frac{4}{2}\ \frac{5}{1}\ \right|\ \cdots$$

They have been grouped according to the sum of the numerator and the denominator of each fraction: first fractions with the sum 2 (there's only one of these), then those with the sum 3 (there are 2), then those with sum 4 (there are 3), then those with sum 5 (there are 4), etc. Now it is easy to set up a one-to-one matching with the natural numbers:

$$\begin{array}{cccccccccccccccc}
\frac{1}{1} & \frac{1}{2} & \frac{2}{1} & \frac{1}{3} & \frac{2}{2} & \frac{3}{1} & \frac{1}{4} & \frac{2}{3} & \frac{3}{2} & \frac{4}{1} & \frac{1}{5} & \frac{2}{4} & \frac{3}{3} & \frac{4}{2} & \frac{5}{1} & \cdots \\
\updownarrow & \updownarrow & \updownarrow & \updownarrow & \updownarrow & \updownarrow & \updownarrow & \updownarrow & \updownarrow & \updownarrow & \updownarrow & \updownarrow & \updownarrow & \updownarrow & \updownarrow & \\
1 & 2 & 3 & 4 & 5 & 6 & 7 & 8 & 9 & 10 & 11 & 12 & 13 & 14 & 15 & \ldots
\end{array}$$

Since it seems intuitively that there are so many more fractions than natural numbers, this demonstration could easily lead one to imagine that every infinite set can be matched up one-to-one with the natural numbers. Cantor's great achievement was to show that this is not the case. The numbers represented by fractions are called *rational.* If a rational number is represented as a decimal, the pattern of digits eventually begins to repeat. Here are some examples:

$$\begin{aligned}
\frac{1}{3} &= 0.33333333333333333333\ldots \\
\frac{1}{4} &= 0.25000000000000000000\ldots \\
\frac{5}{3} &= 1.66666666666666666666\ldots \\
\frac{24}{11} &= 2.18181818181818181818\ldots \\
\frac{9}{7} &= 1.28571428571428571428 57\ldots
\end{aligned}$$

Numbers that can be represented by decimals, whether or not they eventually repeat, are called *real numbers.* Those whose decimal representations never repeat are called *irrational.* Here are some examples of numbers that have been proved to be irrational:

$$\begin{aligned}
\sqrt{2} &= 1.414213562373095050\ldots \\
\sqrt[3]{2} &= 1.259921049894873160\ldots \\
\pi &= 3.141592653589793240\ldots \\
2^{\sqrt{2}} &= 2.665144142690225190\ldots
\end{aligned}$$

Numbers like $\sqrt{2}$ and $\sqrt[3]{2}$ as well as all of the rational numbers are called *algebraic* because they can serve as solutions of algebraic equations. (Thus, $\sqrt{2}$ is a solution of the equation $x^2 = 2$, and $\sqrt[3]{2}$ is a solution of the equation $x^3 = 2$.) The numbers π and $2^{\sqrt{2}}$ have been proved to satisfy no algebraic equation; such numbers are called *transcendental.*

After having shown that the fractions can be matched in a one-to-one manner with the natural numbers, Cantor turned his attention to the set of all algebraic numbers, and he had little difficulty in once again finding a way to match them with the natural numbers in a one-to-one manner. Naturally, he wondered whether the same was true for the set of all real numbers.

We can follow the ruminations of the 28-year-old Cantor in letters written in 1873 to Richard Dedekind, a young mathematician Cantor had met quite by chance the previous year while on vacation in Switzerland. Cantor, who had recently been promoted to a professorship at Halle, wrote showing Dedekind that (as we have already seen), one can construct a one-to-one matching between the natural numbers and the more inclusive set of all positive fractions. He even showed that the same is true for the set of all algebraic numbers. In his letter, Cantor raised the question of the possibility of a one-to-one matching between the set of natural numbers and the set of all real numbers. Dedekind's reply suggested that he believed the question to be of little interest. About a week later, in another letter, Cantor was able to prove to Dedekind the remarkable fact that the set of real numbers *cannot* be matched with the set of natural numbers in a one-to-one manner, that infinite sets come in at least two sizes.

Apparently, Cantor himself wasn't even sure that this finding was worth publishing. He only submitted it for publication after his former teacher Karl Weierstrass encouraged him to do so. The revolutionary implications of what Cantor had done were hardly evident in the four-page paper. The emphasis of the paper was not on the fact that infinite sets had been shown to come in more than one size, but rather that as a corollary, Cantor had obtained a new proof that there exist real numbers that are transcendental. Cantor's proof amounted to noting that since the algebraic numbers can be matched one-one with the natural numbers, and the real numbers cannot be so matched, it follows that the set of real numbers is different from the set of algebraic numbers. It follows that there must be a real number that is not algebraic, that is, transcendental.[11]

Meanwhile, Cantor's personal life flourished. In 1874, he married Vally Guttman, a close friend of his sister and a gifted musician. They had six children and, from all accounts, were a loving devoted family. Although Cantor had a reputation for being forceful and even difficult in professional contexts, he was apparently quite gentle at home. According to one account of mealtime at the Cantor's:

> ...at mealtimes he would sit silently and allow his children to lead the conversation, and then rise and thank his wife for the meal with: "Are you content with me and do you then also love me?"[12]

But as he began devoting more and more of his efforts to developing set theory, Cantor began encountering increasing opposition to his unsettling new ideas. Particularly disappointing was his former teacher Kronecker's unwillingness to accept Cantor's work as a legitimate part of mathematics. In this atmosphere, an appointment to a university where Cantor could have contact with colleagues of his own stature was not to be. He would have to remain in the backwater that was Halle. Cantor's efforts to coax his friend Dedekind to come to Halle failed. In 1886, acquiescing to the inevitable, Cantor purchased a magnificent house for his family in Halle.

Cantor's Quest for Infinite Numbers

Ignoring Gauss's warning that mathematicians had no business with completed infinities, Cantor felt drawn by the lure of the infinite, hitherto the province of theologians and metaphysicians. His mathematical research had provided the basis for his radical ideas, but he pressed on far beyond what that research mandated. The natural numbers $1, 2, 3, \ldots$ are used in ordinary discourse in two different but related ways. They are used to count and to rank, as illustrated by the sentences:

- There are *four* people in this room.
- Joe's horse came in *fourth.*

Everyday language recognizes this with its distinction between *cardinal* and *ordinal* numbers for which different words are provided: *one, two, three, ...* but *first, second, third, ...* Cardinal numbers are used to specify how many things there are in some set; ordinal numbers are used to specify how these things are arranged in a particular order. Cantor's finding that there is no one-to-one correspondence between the natural numbers and the real numbers led him to think about infinite cardinal numbers, and his work on trigonometric series suggested a way to conceptualize infinite ordinal numbers.

Cantor assumed that associated with every set (finite or infinite) there is its unique *cardinal number.* Cantor thought of the cardinal number of a set as obtained by disregarding the specific nature of the items making up the set, so that what remained were simply featureless "units." In particular, if two sets can be matched in a one-to-one manner, then they will have the same cardinal number.

Let M stand for some perfectly arbitrary set. Then Cantor introduced the notation $\overline{\overline{M}}$ for the cardinal number of that set M.[13] For example,* if

$$A = \{\clubsuit \diamondsuit \heartsuit \spadesuit\}, \quad B = \{3, 6, 7, 8\}, \quad \text{and} \quad C = \{6, 5\},$$

*Note the use of curly braces $\{\ldots\}$ to signal that the items listed are thought of as forming a set.

then

$$\overline{\overline{A}} = \overline{\overline{B}} = 4 \quad \text{and} \quad \overline{\overline{C}} = 2.$$

Of course it is easy to set up a one-to-one matching between A and B:

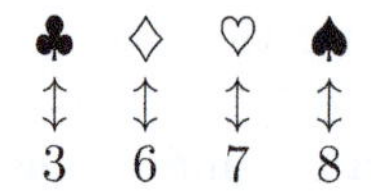

What happens when two sets do not have the same cardinal number? In symbols, it is a question of sets M and N such that $\overline{\overline{M}} \neq \overline{\overline{N}}$. In such a case, one of the two cardinal numbers is the larger and the other the smaller. Using the standard symbols $<$ ("is less than"), and $>$ ("is greater than") we can write $\overline{\overline{M}} < \overline{\overline{N}}$ (or equivalently, $\overline{\overline{N}} > \overline{\overline{M}}$) to indicate that it is N that has the larger cardinal number. To prove that this is indeed the case what is needed is a one-to-one match-up between M and *some subset of* N.[14] Thus, in the example above where $\overline{\overline{A}} > \overline{\overline{C}}$ (because $4 > 2$), the subset $\{\diamondsuit\ \heartsuit\}$ of A can be matched with C like this:

$$\begin{array}{cc} 6 & 5 \\ \updownarrow & \updownarrow \\ \diamondsuit & \heartsuit \end{array}$$

As long as one sticks to finite sets, all of this might well seem to be a matter of expressing simple familiar matters in abstruse terms. Indeed, the power of Cantor's ideas only becomes apparent when applied to infinite sets. Cantor called the cardinal numbers of infinite sets, *transfinite*. His first example of a transfinite number was the cardinal number of the set of natural numbers for which Cantor introduced the symbol $\aleph_0$, usually read "aleph-null," $\aleph$ being the first letter of the Hebrew alphabet.*

Cantor used the symbol $\mathcal{C}$ for the cardinal number of the set of real numbers (because the set of real numbers is sometimes called the *continuum*). Cantor was convinced that $\mathcal{C}$ was the very next transfinite cardinal number after $\aleph_0$. The statement that this is true, that is, that there are no cardinal numbers between $\aleph_0$ and $\mathcal{C}$, is known as Cantor's *continuum hypothesis.* Despite intense efforts over many years, Cantor was never able to resolve the matter: he could neither prove nor disprove the continuum hypothesis. This failure caused Cantor no end of distress. With what we know today, we can see that poor Cantor was just bashing his head against

*Cantor's recourse to the Hebrew alphabet may well be the reason for a rather widespread (and incorrect) assumption that he was a Jew. As Cantor explained in a letter dated April 30, 1895 (my translation), "it seemed to me that for this purpose, other alphabets were [already] over-used." (I'm indebted to Sherman Stein who showed me a copy of this letter.)

a stone wall. Fundamental discoveries by Kurt Gödel in 1938 and Paul Cohen in 1963 revealed that if the continuum hypothesis can be resolved at all, it will require going beyond the methods of ordinary mathematics. So Cantor's inability to settle the matter is hardly surprising. Indeed, even today, experts are divided on the question of whether the Gödel-Cohen negative results are the best that can be expected, or whether new powerful methods may yet yield a more satisfying result.

In Cantor's work on trigonometric series he was led to consider a certain process that could be applied over and over again in stages: a first stage, a second stage, a third stage, and so on. But what pushed Cantor over the edge into the transfinite was the realization that even after all these infinitely many stages, there were more. Soon he was speaking of an ωth stage, an $(\omega + 1)^{th}$ stage, and beyond, and developing the arithmetic of what he came to call *transfinite ordinal numbers.**

Let's look first at the finite set $\{\clubsuit\ \diamondsuit\ \heartsuit\}$. Its members can be ranked in six different ways:

1st	2nd	3rd	1st	2nd	3rd	1st	2nd	3rd	1st	2nd	3rd	1st	2nd	3rd	1st	2nd	3rd
$\downarrow$	$\downarrow$	$\downarrow$	$\downarrow$	$\downarrow$	$\downarrow$	$\downarrow$	$\downarrow$	$\downarrow$	$\downarrow$	$\downarrow$	$\downarrow$	$\downarrow$	$\downarrow$	$\downarrow$	$\downarrow$	$\downarrow$	$\downarrow$
$\clubsuit$	$\diamondsuit$	$\heartsuit$	$\clubsuit$	$\heartsuit$	$\diamondsuit$	$\diamondsuit$	$\clubsuit$	$\heartsuit$	$\diamondsuit$	$\heartsuit$	$\clubsuit$	$\heartsuit$	$\clubsuit$	$\diamondsuit$	$\heartsuit$	$\diamondsuit$	$\clubsuit$

However, these six rankings all exhibit the same pattern: a *first* item, followed by a *second* item, followed by a *third* item. This is true of any finite set: all of the different ways of ranking its members exhibit the same underlying pattern. If the set consists of n items, any ranking will show a first item, a second item, ... and finally an nth item. Cantor saw that with infinite sets the situation is entirely different. Infinite sets can be ranked in different ways with very different patterns. For example, suppose the natural numbers $1, 2, 3, \ldots$ are arranged so that all of the even numbers precede all of the odd numbers like this:

$$2, 4, 6, \ldots, 1, 3, 5 \ldots$$

If we try to use ordinal numbers to show the rank of each item in the progression, we find that the familiar finite ordinal numbers are all used in taking care of the even numbers:

$$\begin{array}{cccccccc} 1^{st} & 2^{nd} & 3^{rd} & \ldots & ? & ? & ? & \ldots \\ \downarrow & \downarrow & \downarrow & & \downarrow & \downarrow & \downarrow & \\ 2 & 4 & 6 & \ldots & 1 & 3 & 5 & \ldots \end{array}$$

*ω is the last letter in the Greek alphabet and is pronounced "omega."

Cantor saw how to use *transfinite ordinal numbers* to handle this difficulty. So after all the finite ordinal numbers, Cantor postulated a first transfinite ordinal number he designated by the Greek letter ω (omega). This was then followed by $\omega + 1$, $\omega + 2$, etc. Cantor could then have provided ranks for the odd numbers in the above example quite easily:

$$\begin{array}{cccccccc} 1^{st} & 2^{nd} & 3^{rd} & \dots & \omega^{th} & (\omega+1)^{th} & (\omega+2)^{th} & \dots \\ \downarrow & \downarrow & \downarrow & & \downarrow & \downarrow & \downarrow & \\ 2 & 4 & 6 & \dots & 1 & 3 & 5 & \dots \end{array}$$

Cantor found that the natural numbers could be ranked in many many different ways using larger and larger transfinite ordinal numbers. He called the finite (ordinal) numbers, that is the natural numbers $1, 2, 3, \dots$, the *first number class* and the transfinite ordinal numbers needed to supply ranks in different arrangements of the natural numbers, the *second number class.* Considering the set of transfinite ordinal numbers constituting this "second number class," Cantor designated its cardinal number by the symbol $\aleph_1$. *Remarkably, Cantor was able to prove that not only is* $\aleph_0$ *the smallest transfinite cardinal number, but that* $\aleph_1$ *is the very next cardinal number after* $\aleph_0$. So, $\aleph_1 > \aleph_0$, and there are no cardinal numbers that are larger than $\aleph_0$ and smaller than $\aleph_1$.

Now, the continuum hypothesis that Cantor had been trying so hard to prove is the statement that $\mathcal{C}$ is the very next cardinal number after $\aleph_0$. Since he knew that $\aleph_1$ really is the next cardinal number after $\aleph_0$, the continuum hypothesis amounted to the succinct question:

$$\mathcal{C} \stackrel{?}{=} \aleph_1.$$

Unfortunately, simply writing this equation brought Cantor no closer to proving that it was true.

After the first and second number classes, is there a third number class? Absolutely! In ranking sets with cardinal number $\aleph_1$ the numbers of the first and second number classes do not suffice. Cantor introduced ω_1 as the transfinite ordinal number beginning the third number class, and he called the cardinal number of the set of all ordinal numbers in this third number class $\aleph_2$. Then, $\aleph_2$ turned out to be the very next cardinal number after $\aleph_1$. Cantor saw that there is no end to this process: after $\aleph_2$ comes $\aleph_3$, then $\aleph_4$, and so on. And after all of these comes $\aleph_\omega$, and on and on.

In developing these ideas Cantor was exploring a domain that had been visited by no one before him. There were no mathematical rules on which he could rely. He had to invent it all himself, relying on his intuition. Considering the nature of the terrain he was investigating, it is remarkable that the bulk of his work has held up very well. But from the beginning there were those who opposed Cantor's entire enterprise. Kronecker's objections have already been mentioned. A story that has been circulating among mathematicians and has been widely believed has it that the influential French mathematician Henri Poincaré said that one day Cantor's set theory "would be regarded as a disease from which one had recovered." Although the story seems to be apocryphal, its very currency shows what Cantor was up against.

The Diagonal Method

If students today learn one thing that Cantor accomplished, it is almost certainly his so-called diagonal method. This method was published in a paper of only four pages in 1891 after Cantor had all but ceased doing mathematical research and after his definitive articles on transfinite numbers had not only been published, but had even been reprinted. It was in 1874 that Cantor had published his proof that there is no one-to-one correspondence between the natural numbers and the real numbers, or, as Cantor would later express it, $\aleph_0 < \mathcal{C}$. The proof used methods borrowed from the basic theory of limit processes as developed by Weierstrass. Using the diagonal method, the same conclusion can be seen to be obtainable from basic logical principles. The diagonal method will come up again and again in our story .

In explaining the diagonal method, it will be helpful to use the metaphor of a labeled package. What will be special is that the things used as labels will be exactly the same sort of thing that is inside the package. As an example, consider the four suits in a deck of playing cards: $\clubsuit\ \diamondsuit\ \heartsuit\ \spadesuit$. Let us use each of them as a label on a package containing some of these same suits, like this:

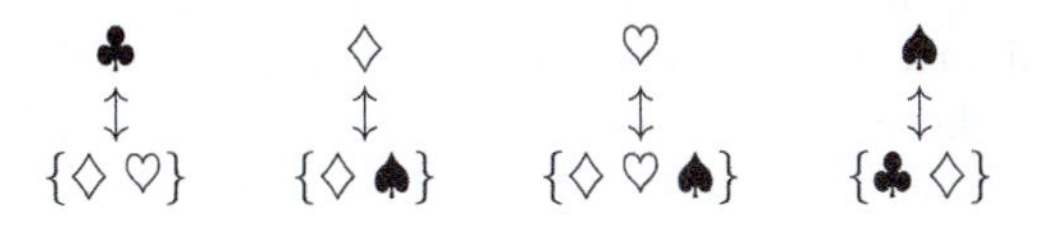

We can exhibit the same information in the form of a table in which we use a plus sign to indicate that an item is inside a package and a minus sign to indicate that it is not:

	$\clubsuit$	$\diamondsuit$	$\heartsuit$	$\spadesuit$
$\clubsuit$	$\ominus$	$+$	$+$	$-$
$\diamondsuit$	$-$	$\oplus$	$-$	$+$
$\heartsuit$	$-$	$+$	$\oplus$	$+$
$\spadesuit$	$+$	$+$	$-$	$\ominus$

In this table, the vertical column on the left lists the four labels, and the contents of the "packages" are displayed in horizontal rows. The plus and minus signs along the *diagonal* are encircled for emphasis. Now the diagonal method is a technique for combining the same kind of item into a new package whose contents is different from each of the labeled packages. Here's how it works: we make a new table in which we insert the opposite sign to the one on the diagonal for each item. Thus, since the sign that goes with $\clubsuit$ is minus, in our new table, $\clubsuit$ gets a plus sign. Similarly $\diamondsuit$ gets minus, $\heartsuit$ gets minus, and finally $\spadesuit$ gets plus, like this:

$\clubsuit$	$\diamondsuit$	$\heartsuit$	$\spadesuit$
+	−	−	+

So, our new package is $\{\clubsuit\ \spadesuit\}$. How can we be sure that it is different from any of the labeled packages? Well, it can't be the package labeled by $\clubsuit$, because $\clubsuit$ *is not* in that package, and it *is* in our new package. It can't be the package labeled by $\diamondsuit$, because $\diamondsuit$ *is* in that package and it *is not* in our new package, and so on.

Now the packages are, of course, sets. And the labeling is a way of setting up a one-to-one matching between the sets and members. The method is perfectly general: it doesn't matter whether you begin with a finite set or an infinite set. If you use each element of that set to label some one particular set made up of some of those same elements, then the diagonal method can be used to obtain a new set of those elements, different from all the sets that have been labeled.

Let's see how this would work if we begin with the set of natural numbers $1, 2, 3, \ldots$. We imagine putting some of these numbers into a package. One package might consist of just the numbers $\{7, 11, 17\}$. Another might consist of all of the even numbers. Now let us imagine using the natural numbers as labels, as in the following infinite array:

$$\begin{array}{ccccc} 1 & 2 & 3 & 4 & \ldots \\ \updownarrow & \updownarrow & \updownarrow & \updownarrow & \ldots \\ M_1 & M_2 & M_3 & M_4 & \ldots \end{array}$$

where each of M_1, M_2, M_3, M_4, $\ldots$ is a package of natural numbers. At this point, we manufacture our new set M different from every one of these by using the following table:

1	– if 1 is in M_1	otherwise +
2	– if 2 is in M_2	otherwise +
3	– if 3 is in M_3	otherwise +
4	– if 4 is in M_4	otherwise +
...	... ...	...

In other words, 1 belongs to M just in case 1 doesn't belong to M_1; 2 belongs to M just in case 2 doesn't belong to M_2; etc. Therefore, M is a set of natural numbers different from M_1, different from M_2, etc. Now because M_1, M_2, M_3, M_4, ... stands for any possible one-to-one matching between the numbers $1, 2, 3, \ldots$ and sets of natural numbers, we see that *no such matching can include all sets of natural numbers.* In other words, the cardinal number of the set of all sets of natural numbers is greater than $\aleph_0$. Actually, it is possible to prove that this cardinal number is none other than $\mathcal{C}$, the cardinal number of the set of real numbers.[15] Thus, the diagonal method provides another way to see that there are more real numbers than natural numbers.

This method is so very general that it provides another way (different from Cantor's successive $\aleph$s) to generate lots of transfinite cardinal numbers. For example, we can think of packages of real numbers labeled by real numbers. The diagonal method shows that no such labeling can include *all* sets of real numbers. Hence the cardinal number of the set of all such sets must be greater than $\mathcal{C}$, the cardinal number of the set of real numbers.[16] And, there is no need to stop there. The question of how cardinal numbers obtained in this manner are interlaced with Cantor's $\aleph_0, \aleph_1, \aleph_2, \ldots$ remains a source of difficulty and controversy to this day.

Depression and Tragedy

From the first Cantor faced opposition based on objections to the very idea that finite human beings living in a finite world could hope to make meaningful assertions about the infinite. But just around the turn of the century, things became much worse with the discovery that unfettered reasoning with Cantor's transfinite could lead to very paradoxical and even ridiculous results.

The trouble all began with attempts to collect the totality of Cantor's transfinite cardinal or ordinal numbers into a single set. If there is a set of all cardinal numbers, what could its cardinal number be? It turned out that it would have to be larger than any cardinal number. How could this be? How could a cardinal number be larger than all cardinal numbers?

Shortly after Cantor became aware of this disconcerting paradox, the Italian mathematician Burali-Forti found a similar difficulty in trying to deal with the set of all transfinite ordinal numbers: he showed that such a set would lead to a transfinite ordinal number larger than any transfinite ordinal number, clearly a ridiculous conclusion.

Then Bertrand Russell came on the scene and delivered the most shocking blow of them all. He considered the question: *can there be a set of all sets?* If there were such a set, what would happen if the diagonal method were applied to it? In other words, what if we thought of packaging up arbitrary sets, and then use sets to label the packages? Of course, we'd get a set different from all those that had been provided with labels. It was in contemplating this situation, that Bertrand Russell found his famous paradox of the set of all those sets not members of themselves. This was the paradox, discussed in the previous chapter, that Russell communicated to the shattered Frege.

Although Russell discovered his paradox by thinking about Cantor's ideas, the paradox itself does not depend at all on considerations involving transfinite numbers. To many mathematicians, it seemed as though the most basic logical reasoning had become unreliable, filled with pitfalls. Not surprisingly, most mathematicians continued their usual work remote from these matters.

But for those who were concerned with fundamental issues about the nature of mathematics, the situation was nothing less than a crisis in the foundations of mathematics. These mathematicians and philosophers soon found themselves dividing into opposing camps. In particular there were those who saw set theory as an integral part of mathematics to be preserved at all costs and those who sought to insulate the body of mathematics from contamination by Cantor's transfinite. Work by logicians during the first three decades of the twentieth century was dominated by these issues.

Cantor suffered the first in a series of "nervous breakdowns" in 1884, an intense depression that lasted about two months. Having recovered, Cantor attributed his mental problems to the intensity of his work on the continuum hypothesis and to Kronecker's unfavorable view of his work. At this time he even wrote Kronecker a letter proposing that they renew a friendly relationship to which Kronecker responded cordially.

Cantor's explanation of his difficulties was widely accepted for many years despite several episodes of what is now viewed as bipolar disease. Regardless of the severity of external events, it is now generally understood that the disorder's fundamental cause is rooted in defective brain chemistry, with environmental factors, like those to which Cantor had attributed his depression, precipitating rather than causing major disruptions.[17]

This episode pretty much marked the end of Cantor's ground-breaking work in the theory of sets except for the paper already mentioned on the diagonal method. More and more, between episodes of severe mental illness, Cantor concerned himself with philosophy, theology, and, of all things, the question of the authorship of Shakespeare's plays. About this last topic, Cantor developed notions of the significance of this work and of those eager to suppress it that bordered on paranoia. The year 1899 brought crisis and tragedy for Cantor. It was the year in which he first faced the paradoxes of set theory. And then he suffered a devastating loss with the death of his beloved 13-year-old son.

It was not Georg Cantor's style to pursue a subject as a dilettante. He made himself an expert on the Elizabethan period in general and on Shakespeare's plays in particular, and published a number of monographs purporting to prove that Shakespeare's plays had been written by Francis Bacon. This, of course had nothing to do with set theory or the transfinite. However, Cantor saw his investigations in philosophy and theology as definitely connected with his work on the infinite. Cantor believed that beyond the transfinite there is an "absolute infinite" that mere human understanding can never fully encompass. Even the bedeviling paradoxes that arose in set theory were to be understood from this point of view. For example, the multitude of all transfinite cardinals should be regarded as being absolutely infinite, and that is why contradictions arise from thinking of them as merely transfinite.

A Decisive Battle?

In German philosophical thought, the towering figure was Immanuel Kant whose critical philosophy was framed by two key questions:

- How is pure mathematics possible?
- How is pure natural science possible?

Kant's answer to the first question relied on what he called "pure intuitions" of space (for geometry) and of time (for arithmetic). He conceived of these intuitions as being entirely independent of empirical sensations.[18] Despite Kant's emphasis on the importance of science, post-Kant philosophy in nineteenth-century Germany evolved in a different direction, moving to an "absolute idealism" that conceived of *ideas* and *concepts* as primary and sought to understand the world almost as though these were what it was made of.

One of the leaders of this movement was Georg Wilhelm Friedrich Hegel whose lectures were attended by hundreds of eager disciples. Hegel had many followers (among whom, famously, were Karl Marx and Friedrich Engels), and scholars still find much worthwhile in his writings. However, he was capable of contorted reasoning that simply invites ridicule, especially in his massive two-volume *Science of Logic* in which readers were asked to ponder the deep thoughts:

> *Nothing* is simple equality with itself.
>
> *Being* is *Nothing.*
>
> *Nothing* is *Being.*
>
> Both of these categories in the transition from each to the other dissolve into the further category: *Becoming.*

Meanwhile, towards the end of the century, deriving its impetus in part from the "positivistic" ideas of August Comte, and partly from developments in science, a new "empiricist" philosophy was developing in Germany. For the empiricists, the primary items in terms of which the world is to be understood are sense data. Cantor saw this empiricism as a reaction to nonsense like Hegel's, but found it to be crude and simplistic.

The great scientist Hermann von Helmholtz, one of the principal exponents of empiricist philosophy, wished to bring back Kant's central focus on empirical science. A little pamphlet he wrote on counting and measuring infuriated Georg Cantor. In 1887, in an article surveying transfinite numbers from mathematical, philosophical, and theological viewpoints, Cantor made a point of attacking this pamphlet as expressing an "extreme empirical-psychological point of view with a dogmatism one would not have thought possible . . . " He went on to complain:

> Thus, in today's Germany we see, as a reaction against the overblown Kant-Fichte-Hegel-Schelling Idealism, an *academic-positivistic skepticism* that powerfully dominates the scene. This skepticism has inevitably extended its reach even to *arithmetic*, in which domain it has led to its most fateful conclusions. Ultimately, this may turn out most damaging to this positivistic skepticism itself.

This article was included in a collection of Cantor's papers dealing with the transfinite published in 1890. Frege, given the task of reviewing the volume, chose to emphasize the remark just quoted. In a remarkable passage (already quoted in part at the beginning of this chapter), appearing in print just a decade before he was to receive Bertrand Russell's devastating letter, Frege wrote:

> Yes indeed! This is the very reef on which it will founder. For ultimately, the role of the infinite in arithmetic is not to be denied; yet, on the other hand, there is no way it can coexist with this epistemological tendency. Thus we can foresee that this issue will provide the setting for a momentous and decisive battle.[19]

Georg Cantor died suddenly of heart failure on January 6, 1918, while the World War was still raging. Today although the battle that Frege predicted in his military metaphor has provided many surprises, it has hardly resulted in any decisive outcome. Perhaps the most surprising by-product of this battle was Alan Turing's mathematical model of an all-purpose computer.

Appendix: Cantor and Kronecker

From school mathematics, one easily gets the impression that mathematics is a stale subject that had reached its finished form long ago. It was largely through the writings of E.T. Bell that, as a teenager, I came to understand that mathematics is an exciting activity in which the answers to open questions are eagerly sought by professional mathematicians. Bell, a professor of mathematics at the California Institute of Technology, was a prolific writer. In addition to his popular books on mathematics and mathematicians, he was the author of over 200 technical papers, as well as a number of science fiction novels written under the name John Taine. One could with justice say, however, that Bell never let the facts stand in the way of a juicy story. In his still popular *Men of Mathematics,* he wrote as follows:

> Rightly or wrongly, Cantor blamed Kronecker ... The aggressive clanishness of Jews has often been remarked, sometimes as an argument against employing them in academic work, but it has not been so generally observed that there is no more vicious academic hatred than that of one Jew for another when they disagree on purely scientific matters ... When two intellectual Jews fall out they disagree thoroughly, throw reserve to the dogs, and do everything in their power to cut one another's throats.[20]

This passage is vile in so many ways that one's first reaction might well be to not dignify it by discussing it. But it must be discussed because this and other things that Bell wrote about Kronecker and Cantor have had an enduring effect. To begin with, it is appalling that a distinguished scholar would resort to such prejudiced stereotypes. Especially in 1937, when Hitler and the Nazis had been in power in Germany for more than four years, and Bell's Jewish colleagues there were dismissed from their positions and otherwise persecuted,

Bell's comment showed a hard-to-believe callous insensitivity. But what is worse, it isn't even true! Kronecker was indeed a Jew and readily affirmed the fact. Cantor may well have had one or more Jewish ancestors, but his family had been Christian for two generations, and Cantor himself had a deep interest in Christian theology. In a later edition of the book, the antisemitism has been removed, but we still can read "When Cantor and Kronecker fell out they disagreed all over, ... and did everything but slit the other's throat."[21]

Bell continues the theme of Kronecker as Cantor's implacable enemy. Cantor's teachers during his student days included "his future enemy Kronecker."[22] Kronecker was not merely Cantor's enemy, he was "his arch-enemy."[23] Bell certainly doesn't hesitate to show us Cantor's bloody wounds: "Kronecker attacked 'the positive theory of infinity' and its hypersensitive author vigorously and viciously with every weapon that came to hand, ..."[24] Colorful stories about famous mathematicians are passed on as gossip from one generation of mathematicians to the next. Quite possibly Bell had picked up these stories in that way. But even in a book intended for a general audience, responsible scholars owe it to their readers to be reasonably sure that what they write is true.

I must confess that in previous editions of this book, I uncritically followed the crowd in accepting this account of the relationship between Cantor and Kronecker without making any effort to verify any of it.

Kronecker, a great mathematician, had very decided views that found him opposed to the mathematical ideas developed by his colleagues Cantor, Dedekind, and Weierstrass. He didn't believe that completed infinities had any legitimate place in mathematics. He was a virtuoso practitioner of his kind of algorithmic mathematics based on the arithmetic of natural numbers, and he insisted that assertions that some mathematical object exists are meaningless without an algorithm for calculating that object.[25] He certainly did not like Cantor's mathematics and was not shy about expressing his opinion.

There is scant evidence of any personal animosity,[26] but lots of evidence of Cantor's ambivalent attitude towards Kronecker. He wrote many letters in which he expressed a belief that Kronecker was acting to interfere with the publication of his work. At the same time he seems to have had friendly relations with Kronecker, being a guest at his house in Berlin on more than one occasion. It should be realized that, given attitudes in Europe at that time, in relation to one of his teachers, Cantor would have felt that etiquette required him to show a certain deference. One of these visits followed a letter from Cantor to Kronecker. In the letter Cantor wrote:*

> As a result of a certain sharpness in the evaluation of my scientific works, I have come to be, not without some guilt myself in the matter, in a position of opposition to you from which I yearn to extricate myself. In the letter Cantor also said, "Perhaps, by giving a more accurate explanation of my works ..., I

*I am indebted to my good friend and colleague Harold Edwards, an expert on Kronecker's mathematics and a distinguished historian of mathematics, for calling my attention to the lack of evidence for the common view of Kronecker as Cantor's implacable enemy. He graciously made his extensive research on this matter available to me.

> will succeed in causing you to be better disposed toward them, . . . "

Kronecker replied:

> I have just received your kind note . . . and . . . I have received it with gratitude . . . and I would be very happy if you . . . would visit me . . . and we could have a scientific discussion, as we so often have in the past.

In the letter Kronecker noted a "divergence" in their scientific views, but said, "I see no reason whatever that our personal relations should be disturbed by this divergence." The letter was signed "your old friend, Kronecker"[27] This meeting did occur and, although Cantor had no success in persuading Kronecker to change his views, it seems to have been quite amicable. After the meeting Cantor wrote to a friend as follows:

> I arrived at seven o'clock in the evening to have a visit. However, he invited me to remain for tea en famille which I accepted. The discussions lasted until 1AM. . . . As far as my personal relations with Kronecker are concerned, they are and remain excellent after I approached him in the most conciliatory mannner, and he most cordially accepted the hand I had proffered.[28]

This hardly accords with Bell's picture of Kronecker and Cantor doing "everything in their power to cut one another's throats."

What Bell wrote would have little continuing significance, but, alas, much of it has come to be generally accepted. In previous editions of this book I called Kronecker "Cantor's nemesis," and I was hardly alone. And in recent years, a scholar with a sterling reputation, Joseph Dauben, Professor of History at the City University of New York and winner of prestigious awards, in his writings affirms the picture Bell had drawn and provides it with a scholarly patina.

I am indebted to his book on Cantor's ideas for calling my attention to Frege's prediction of a "momentous and decisive battle" quoted in the previous section. However, when I read what Dauben said, it was clear that he could not have understood Frege. Dauben wrote: "This was the issue, said Frege, over which mathematics will be wrecked." No one who had any understanding of Frege's work and ideas could imagine that in 1892, he would have believed that mathematics was in danger of being "wrecked." He quotes from Frege correctly: "Here is the reef on which it will founder." But he seems to suppose that the pronoun "it" ("sie" in the German original) refers to mathematics.

This is very strange because this pronoun must refer to something in the passage from Cantor to which Frege is responding, and that passage contains no reference to the subject of mathematics as a whole. It is clear from the context that the pronoun refers to "positivistic skepticism." Frege agrees with Cantor that this positivistic skepticism can not provide what is essential for the philosophy of arithmetic and suggests that it will be this very positivistic skepticism that will be damaged in the encounter.[29]

Dauben continues his theme of Frege as prophesizing that mathematics would be "wrecked." In "The Splintering of Mathemtics," he writes:

> Frege's fatal prediction of 1892 concerning the future of mathematics had come true in little more than a decade.

Dauben then presents the great French mathematician Poincaré as an implacable opponent of Cantor's mathematics:

> Intuitionists like Poincaré argued that most of the ideas of Cantorian set theory should be banished from mathematics once and for all. ... Transfinite set theory, Cantor's great contribution to mathematics, involved nothing in Poincaré's view but contradictory and therefore meaningless concepts. The paradoxes of set theory were direct evidence that Cantor's ideas were a grave disease that seemed to infect all mathematics. Poincaré's medicine was hard stuff indeed: his prescription called for the elimination of virtually every aspect of Cantor's work from respectable, permissible, and finite mathematics.[30]

In support of this, Dauben cites Poincaré's address titled "The Future of Mathematics," delivered to the International Congress of Mathematicians in Rome in 1908. Poincaré's skeptical attitude toward Cantor's work has been noted earlier in this chapter. What he said on this occasion was much milder and in no way justifies Dauben's assertions. In a brief paragraph on "Cantorism," Poincaré said:

> I spoke earlier of the need to return again and again to the first principles of our science and of the gains this could yield for the study of the human mind. It is this need that has inspired two efforts that have held a very large place in the most recent history of mathematics. The first is Cantorism which has rendered to science the services that we all know. One of the characteristic features of Cantorism is this: instead of achieving generality by building up more and more complicated constructions and defining by construction, it proceeds by beginning with a sufficiently large set, and defines only by specifying a concept as an appropriate subset. Hence the horror that it has at times engendered in some minds, Hermite's for example, whose preferred stance was to compare mathematical science to natural science.

> For most of us these prejudices had dissipated, but then we came up against paradoxes, apparent contradictions that would have elated Zeno of Elea and the Megara school. And now everyone is busy seeking the remedy. Speaking for myself, and I am not alone, I believe it is important to only introduce entities that can be completely defined in a finite number of words. Whatever will be the remedy that will eventually be adopted, we can look forward to the pleasure of the physician asked to follow a beautiful pathological case.[31]

This passage is certainly not friendly to Cantor's work, but it hardly supports Dauben's assertions. It is clear that Poincaré is using the word "Cantorism" to include much more than Cantor's own work. In fact, Cantor's transfinite numbers are not even mentioned. The reference to nonconstructive proofs likely refers to work on the foundations of calculus by Weierstrass and Dedekind as well as Cantor, and perhaps also on Hilbert's work on invariants. (Hilbert is the subject of the following chapter, in which his work on invariants will be discussed.).

Hermite's notorious "horror" was over an example that Weierstrass had produced. And rather than calling for the Cantorian ideas to be extirpated from mathematics, as Dauben insists, Poincaré says that mathematicians had become used to them and it was only the paradoxes that were now a problem. And the medical metaphor suggests, not that "Cantorism" is a disease infecting the body politic of mathematics, but rather that a cure to the paradoxes would be found and would yield great pleasure.

Finally, if Dauben had perused the proceedings of the 1908 Congress, rather than a "wrecked" splintered mathematics, he would have found a flourishing enterprise with addresses given in four languages on diverse topics. Other than Poincaré's address, which relegated his discussion of foundational matters to the final page of a 15-page essay on the future of mathematics, none of the plenary addresses dealt with foundational matters.

From other parts of Poincaré's address, it is apparent that he had no intention of abandoning the services that Cantorism had rendered.[32] Kronecker, on the other hand, rejected these services. He maintained that one should always define "by construction," and that working with completed infinities was both unnecessary and of dubious validity. Whether Kronecker, had he not died almost two decades before the 1908 Congress, would have eventually come to see the benefits of Cantorism to mathematics is something we will never know.

After my realization that Dauben had seriously misunderstood what Frege had said, I should have shown some skepticism toward what he said about Kronecker and Cantor. But I did not. In previous editions of this book, basing myself on Dauben's book, I wrote that Kronecker had tried to prevent the publication of one of Cantor's early papers on set theory. The paper had been submitted to a journal whose editorial board included Kronecker. Indeed, in his book Dauben asserted:

> Though Kronecker did not succeed in suppressing Cantor's paper, the delay in publication represented the first major conflict Cantor was to experience over the acceptance of his work. [This] was the first occasion for open hostility.

In a more recent article entitled *The Battle for Cantorian Set Theory*, concerning this same paper, Dauben wrote, "Kronecker ... delayed publication of a paper Cantor had written."[33] The only evidence Dauben provides in either publication is a letter Cantor wrote in which he expressed the belief that his paper was not being published promptly. Grattan-Guiness, a reputable and distinguished historian of mathematics wrote:

> Allegedly Kronecker had held up publication of this paper ... Cantor himself is the principal source of this story, ... In fact, if there was a delay, it cannot have been a long one (the date of submission of the paper, 11 July 1877, is not obviously out of line with others in the same volume); and, given the way in which Cantor had chosen to express himself, Kronecker deserves our sympathy.[34]

What can we conclude from all of this? To begin with, that something can be generally believed and be asserted by reputable experts, and still be false. As far as Kronecker and Cantor are concerned, what evidence there is about their relationship almost all comes from letters written by Cantor.

Cantor was a brilliant innovative thinker who, based on a few abstract principles and his sound mathematical intuition, created a new mathematics of the infinite. His theorems have survived intact and continue to be developed by set theorists whose efforts are grounded in a much more rigorous framework than was available to him.

Cantor was a deeply troubled man, afflicted by the ups and downs of bipolar disease. He was much distressed by the fact that his teacher, Kronecker, apparently regarded his work as pointless. He believed at various times that Kronecker's opposition went beyond mere words, that he acted to delay or suppress his work. However, his personal relations with Kronecker remained cordial, and all of his contributions were published despite their radical content. Kronecker, a wealthy man, could feel aloof from academic squabbles, content to pursue his mathematics, making immense contributions, using the methods he found appropriate. And we can admire both men and, to the extent that our interest, ability. and training permits, enjoy the fruits of their labors.

CHAPTER 5

Hilbert to the Rescue

In 1737, George II of England, son of Leibniz's last patron George I, founded a university in the medieval town of Göttingen located on the Leine River in central Germany. The city walls, several Gothic churches, and half-timbered houses on old streets survive to this day in this charming university town. Göttingen University has a proud tradition of mathematical excellence dating back to the nineteenth century, having been home to such mathematical greats as Carl Friedrich Gauss, Bernhard Riemann, Lejeune Dirichlet, and Felix Klein. But the true glory days for mathematics in Göttingen came in the twentieth century when, drawn mainly by David Hilbert's reputation, students from everywhere came to what remained the undisputed world center for mathematics until the exodus resulting from the Nazi takeover of Germany in 1933.

During my graduate student days in the late 1940s, anecdotes about Göttingen in the 1920s were repeated from one generation of students to the next. We heard about the endless cruel pranks that were played on the mathematician Bessel-Hagen who remained ever gullible. My own favorite story was about the time that Hilbert was seen day after day in torn trousers, a source of embarrassment to many. The task of tactfully informing Hilbert of the situation was delegated to his assistant, Richard Courant. Knowing the pleasure Hilbert took in strolls in the countryside while talking mathematics, Courant invited him for a walk.

Courant managed matters so that the pair walked through some thorny bushes, at which point Courant informed Hilbert that he had evidently torn his pants on one of the bushes. "Oh no," Hilbert replied, "they've been that way for weeks, but nobody notices." It was during these same 1920s that Hilbert mounted a remarkable campaign to use mathematics to validate itself. It was a strange sequence of events that led from Hilbert's campaign to Alan Turing's insight into the nature of computation.

David Hilbert was born and grew up in a Protestant family in the town of Königsberg in the Eastern part of Prussia, a town proud to remember that it had been the home of the philosopher Immanuel Kant. In 1870 when Bismarck orchestrated a war with the France of Napoleon III, and used the

DAVID HILBERT
(Author's Collection)

stunning German victory to accomplish the unification of Germany into an empire with the King of Prussia as its Kaiser, Hilbert was a child of eight. By the time he entered the University of Königsberg to study mathematics, his remarkable talent for the subject had been recognized, and his characteristic style of absorbing mathematics in conversation established. With his friends Hermann Minkowski and Adolf Hurwitz, he would go on long walks, talking mathematics all the way.[1]

During the two centuries that had elapsed between the invention of the calculus by Leibniz and Newton and the years when David Hilbert was becoming a mathematician, a host of workers had found many spectacular applications of limit processes. Many of these results were obtained by purely formal manipulation of symbols with little concern for their underlying meaning. But by the middle of the nineteenth century, a day of reckoning had arrived. Problems were arising that demanded conceptual understanding that went beyond mere symbols. At the forefront of this effort were Georg Cantor, his teacher Karl Weierstrass, and his friend Richard Dedekind.

In 1888 Hilbert went on a trip to the major centers of mathematics in Germany to make contact with the leading figures in his field. In Berlin he visited Leopold Kronecker, renewing an acquaintance he had made two years earlier. Kronecker was a great mathematician, some of whose work was to play a fundamental role in Hilbert's accomplishments. But, as Hermann Weyl, Hilbert's one-time student, wrote in an obituary notice half a century later, Hilbert saw Kronecker as using "his power and authority to stretch mathematics upon the Procrustean bed of arbitrary philosophical principles ..."

These "principles" led Kronecker to a profoundly negative attitude towards a good deal of the mathematics of his day. It was not only Cantor's transfinite that Kronecker found objectionable, but also the entire effort by Weierstrass, Cantor, and Dedekind to provide a firm rigorous foundation for the limit processes of the calculus. Kronecker felt that these efforts were unnecessary and unreliable. He was particularly insistent that mathematical proofs of existence be *constructive*. That is, to be acceptable to Kronecker, a proof that there actually are mathematical entities satisfying certain conditions would have to provide a method to explicitly exhibit the entities in question. Hilbert would soon challenge this dictum in his own work. Many years later, he would explain the distinction to students by pointing out the certainty that among the students in the lecture hall (none of whom, apparently, was totally bald), there was one with the least number of hairs on his head although he had no evident way to identify such a student.[2]

Hilbert's Early Triumphs

The world is in flux, but some things do not change. Mathematicians are often concerned with finding out exactly which things stay the same when other things change. In such a case, they speak of things that are *invariant* under certain *transformations.* The investigation of what came to be called *algebraic invariants* was initiated by George Boole in one of his early papers.[3] By the final quarter of the nineteenth century, algebraic invariants had become a major focus of mathematical research. Heroic bouts of algebraic manipulation were brought to bear on the problem of finding invariants.

A true virtuoso in this endeavor was the German mathematician Paul Gordan, dubbed by his contemporaries the "king of invariants." Threading his way through thickets of algebra, Gordan was led to conjecture a simplifying theorem about the structure of algebraic invariants. According to Gordan's conjecture, in considering all of the invariants of a particular algebraic expression, there would always be a finite number of key invariants in terms of which all of the others could be expressed by means of a simple formula. However, his direct onslaught enabled him to prove his conjecture only in a very special case.

Gordan's conjecture was regarded as one of the major problems faced by mathematicians of the day, and it was generally supposed that the person who managed to prove it would do so by displaying a virtuosity with manipulative algebra rivaling Gordan's. In this climate, David Hilbert's proof of Gordan's conjecture came as a great shock. Instead of complicated formal manipulations, Hilbert relied on the power of abstract thought.

It was after meeting Gordan that Hilbert found himself captivated by the problem Gordan had set. His solution, found after six months of work, rested on an extremely general result, known today as *Hilbert's Basis Theorem,* whose proof was quite straightforward. Using this theorem, Hilbert demonstrated that the supposition that Gordan's conjecture is false leads to a contradiction.

This spectacular proof of Gordan's conjecture would not have been satisfactory to Kronecker because of its non-constructive nature. Instead of furnishing a list of the key invariants whose existence Hilbert had established, this proof had merely shown that the supposition of their nonexistence would lead to a contradiction. However, with its demonstration of the power of abstract thought, Hilbert's proof opened a window on the mathematics of the coming century.

The more general viewpoint uncovered by Hilbert's proof had the incidental effect of killing the classical theory of algebraic invariants. Today, Gordan is mainly remembered for his reaction to Hilbert's proof. "This is not mathematics," he exclaimed, "it is theology."

After the sensation created by his solution of Gordan's problem, which had immediately elevated him to the first rank among contemporary mathematicians, Hilbert did not rest on his laurels. But before leaving the theory of invariants for good, he cleaned up some details, in particular giving another proof of Gordan's conjecture that was fully constructive.[4] In addition he published a veritable barrage of papers on a variety of mathematical topics.

Strikingly, one short paper had an unmistakably "Cantorian" flavor that Kronecker surely would have disdained. However, notwithstanding his prolific output, Hilbert's practical career continued to languish as he remained for years a Privatdozent in Königsberg, dependent for his meager earnings on fees for his lectures. In one case, he delivered an entire course of lectures to just one student, a mathematician from Baltimore. In a letter to his good friend Minkowski, Hilbert remarked ironically that there were 11 dozents competing for the same number of students.

The year 1892 marked some crucial changes in young Hilbert's life. It all began with the death of the 68-year-old Kronecker just before the new year, and the retirement of Karl Weierstrass. The closed world of academic mathematical life in Germany began to unfreeze, leading to a virtual game of academic musical chairs in German mathematics. At last, after six years as a Dozent, Hilbert could finally move into a regular academic position at Königsberg.

In this same year he married Käthe Jerosch, his favorite dancing partner. A year later, his son Franz was born. Meanwhile, Felix Klein, the leading light of the mathematics faculty at Göttingen, was determined to lure Hilbert there. By the spring of 1895, Klein's maneuverings had proved successful, and Hilbert had moved to Göttingen where he remained until his death 48 years later.

If Hilbert's dazzling proof of Gordan's conjecture had brought closure to the classical theory of algebraic invariants, his comprehensive *Zahlbericht* (literally, "report on number"), produced at the behest of the German Mathematical Society, was an opening onto a vast mathematical panorama. The society had expected a report on the current state of a relatively new branch of mathematics, algebraic number theory, a topic many mathematicians had been finding baffling.[5] What they got was a critically thought-out reworking of the field from first principles. We were still studying it with pleasure and profit in my graduate-student days half a century later.

Hilbert had come to Göttingen with lectures already prepared for courses on a great variety of mathematical subjects, because of the lectures he had been giving during his dozent days in Königsberg. Otto Blumenthal, the first of the 69 students to complete a doctoral dissertation under his supervision, was able to report 40 years later his clear recollection of the impression Hilbert made on him on when he arrived in Göttingen: "This medium-sized, nimble man with his broad reddish beard and his quite ordinary clothes seemed quite unprofessorial ... [in comparison to the other professors]."

Blumenthal describes Hilbert's lectures as "very much to the point, but with a rather dull delivery style and a tendency to repeat important propositions. However, the rich content and the simple clarity of the presentation led one to forget the form. He would introduce things that were new and that he himself had done, without making a special point of it. It was evident that he took pains to see that everyone understood; he lectured for the students, not for himself."[6]

Students were astonished to find that for the winter 1898 term, Hilbert was proposing to give a course entitled *Elements of Euclidean Geometry.* They had thought of him as being entirely immersed in algebraic number theory and had no notion that he might be interested in geometric subjects. The topic announced seemed particularly strange because Euclidean geometry was, after all, a subject in the secondary-school curriculum. Astonishment only grew when the course began and the students found themselves exposed to an entirely new development of the *foundations* of geometry. This was the first hint of Hilbert's profound interest in the *foundations of mathematics.* It is this interest that will provide our main focus.

In his lectures, Hilbert provided a set of axioms for geometry that plugged some gaping holes in Euclid's classic treatment. He emphasized the abstract nature of the subject: it must be shown by pure logic that the theorems follow from the axioms without the corrupting influence of what we can "see" by looking at a diagram. In a famous anecdote, he is alleged to have said that the theorems must continue to hold if, instead of points, lines, and planes, one were to talk of "tables, chairs, and beer mugs" so long as these latter objects are assumed to obey the axioms.

Finally, to put a cap on his achievement, Hilbert provided a proof that his axioms are consistent, that no contradiction can be derived from them. This proof showed that any inconsistency in his axiom system for geometry would result in an inconsistency in the arithmetic of real numbers. So what Hilbert had done was to *reduce* the consistency of Euclidean geometry to that of arithmetic, leaving the problem of the consistency of arithmetic for another day!

Towards a New Century

The mathematicians present at an international conference in Paris in August 1900 inevitably wondered what the new century would bring to their subject. It was on a sultry day that the 38-year-old David Hilbert, whose stunning accomplishments had taken him to the top of his profession, was delivering an invited address in which he presented, as a challenge to the mathematicians of the twentieth century, 23 problems that seemed utterly inaccessible by the methods available at the time.[7] In a burst of characteristic optimism, Hilbert declared that every mathematician shares the conviction

> "that every definite mathematical problem must necessarily be susceptible of an exact settlement ... This conviction ... is a powerful spur to our work. We hear within us the perpetual call: *There is the problem. Seek its solution. You can find it by pure reason ...*"

The first problem on Hilbert's list was deciding the truth of Cantor's continuum hypothesis (the assertion that there are no sets with a cardinal number between that of the set of natural numbers and that of the set of all sets of natural numbers). This was a ringing endorsement of Cantor's transfinite despite the apparent threat posed by paradoxes.

The second problem was precisely the loose end left by Hilbert's proof of the consistency of the axioms of Euclidean geometry: to somehow establish the consistency of the axioms for the arithmetic of real numbers. Previous consistency proofs had been proofs of *relative consistency*; this means that they had worked by reducing the consistency of one set of axioms to that of another. But Hilbert realized that with arithmetic he had reached logical bedrock, and new "direct" methods would be needed.

This problem also provided an opportunity for Hilbert to explain his own view of the meaning of *existence* in mathematics. Whereas Kronecker had proclaimed that to prove the existence of mathematical entities requires that a method be provided for "constructing" or "exhibiting" the items in question, for Hilbert existence simply required a proof that assuming the existence of such entities would not lead to a contradiction:

> "... if it can be proved that the attributes assigned to a concept can never lead to a contradiction by the application of a finite number of logical processes, I say that the mathematical existence of the concept ... is thereby proved."

According to Hilbert, the contradiction arising from supposing the existence of a set consisting of all of Cantor's transfinite cardinal numbers merely showed that such a set does not exist. Especially after the paradox that Bertrand Russell communicated to Frege in his devastating letter of 1902 became generally known, the difficulties with the foundations of mathematics began to be seen as constituting a crisis, and the problem of the consistency of arithmetic continued to fester. It was only during the 1920s that Hilbert with his students and disciples launched a frontal attack on this problem with consequences they could hardly have foreseen.

The set of 23 problems that Hilbert proposed in 1900 has fascinated generations of mathematicians. The problems spanned a great variety of topics in pure and applied mathematics, and presaged the breadth of Hilbert's own contributions to come. In his obituary essay about Hilbert, Hermann Weyl commented that anyone who had solved one of the problems on Hilbert's list thereby entered "the honors class of the mathematical community."

In 1974, the American Mathematical Society sponsored a special symposium (I was privileged to be a participant) in which experts were invited to speak on the mathematical developments that had arisen from these problems in the intervening years. The fecundity of the Hilbert problems can be seen in the fact that the proceedings of this symposium were published in a volume of over 600 pages.[8]

The Battle over the Infinite

The misgivings many mathematicians felt about Cantor's transfinite, and indeed about the entire direction of foundational research, came to a head with Bertrand Russell's making known the contradiction he had found in what seemed to be straightforward reasoning. As we have seen, Frege simply gave up on his life's work when he received a letter containing Russell's paradox. One may wonder whether Frege recalled the prophecy he had made ten years earlier:

> For ultimately, the role of the infinite in arithmetic is not to be denied; . . . Thus we can foresee that this issue will provide the setting for a momentous and decisive battle.[9]

Although Frege and Cantor's friend Dedekind withdrew from the battle, there was no lack of warriors to enter the fray. In the early years of the twentieth century, Hilbert and Henri Poincaré were generally thought to be the two greatest living mathematicians, and they both engaged with gusto, but on opposite sides.

After 1900, the next International Congress of Mathematicians occurred in 1904, two years after Russell had announced his paradox. In Hilbert's address to the congress, he made evident his approach to the "crisis" by outlining the form a consistency proof for arithmetic might take.[10] He did not fail to add that the proof could be extended to encompass Cantor's transfinite as well.

Poincaré was quick to observe that Hilbert was guilty of circular reasoning: the very methods the proof was intended to justify were used in the supposed proof that those methods cannot lead to a contradiction. It would be some years before Hilbert came to terms with this objection. Poincaré saw some use in what he called "Cantorism"; however, he insisted:

> *There is no actual (given complete) infinity.* [Poincaré's italics] The Cantorians have forgotten this, and they have fallen into contradiction.[11]

Here Poincaré is echoing the words of Gauss written some eight decades earlier, already quoted in the previous chapter: "I protest above all against the use of an infinite quantity as a *completed* one, which in mathematics is never allowed." Cantor's great life work had been a heroic challenge to this tradition.

Bertrand Russell was not one of those who retired from the battlefield. He worked assiduously to develop a system of symbolic logic in terms of which Frege's project to reduce arithmetic to pure logic could be carried out without running into the paradoxes. In communicating his efforts to his contemporaries, he was aided greatly by the symbolization introduced by the Italian logician Giuseppe Peano (essentially the one introduced in Chapter 3), far easier to penetrate than Frege's. Poincaré bitterly attacked Russell's efforts:

> It is difficult to see that the word *if* acquires when written $\supset$, a virtue it did not possess when written *if*.

Nor did Poincaré fail to note that to take Russell's effort seriously would open the possibility of reducing mathematics to mere computation (Leibniz's dream!), and to ridicule the very idea:

> Thus it will be readily understood that in order to demonstrate a theorem, it is not necessary or even useful to know what it means. ... we might imagine a machine where we should put in axioms at one end and take out theorems at the other, like that legendary machine in Chicago where pigs go in alive and come out transformed into hams and sausages. It is no more necessary for the mathematician than it is for these machines to know what he is doing.

Bertrand Russell's effort to resurrect Frege's program took the form of the monumental three-volume *Principia Mathematica* (published in the years 1910–1913) that Russell authored with Alfred North Whitehead. This work started out with the pure logic of Frege's *Begriffsschrift* and ended with subject matter that was clearly mathematics with simple direct steps in between, very much in the spirit of Poincaré's Chicago machine. The paradoxes were avoided by an elaborate and unwieldy structure of layers in which, in effect, any particular set could only have members from just one layer. This layering so crippled ordinary mathematics that a special dubious *axiom of reducibility* was provided to cut through the fences that had been erected between the layers.[12]

The *Principia* was also marred by an underlying confusion. While Frege had understood clearly that he was dealing with two levels of language, a new formal language he was constructing and ordinary language in which this new language could be discussed, the Whitehead-Russell opus was unclear on this matter and commingled the two levels.[13] This meant that the problem of the consistency of the entire structure, so crucial for Hilbert, would not even arise in Russell's context. Despite all this, the *Principia* was a landmark achievement demonstrating once and for all that the complete formalization of mathematics in a system of symbolic logic is perfectly feasible.

While Bertrand Russell labored to find a logical basis for the full breadth of classical mathematics while avoiding the paradoxes, a brilliant young Dutch mathematician, L. E. J. Brouwer had convinced himself that much of it was fatally flawed and needed to be discarded. Brouwer's doctoral dissertation of 1907 showed great hostility to Cantor's transfinite and to much of contemporary mathematical practice. In 1905, Brouwer had taken time from his mathematical pursuits to publish a short book, *Life, Art and Mysticism,* drenched in romantic pessimism. After portraying life in this "sad world" as an illusion, this morose young man concluded with:

> Look at this world, full of wretched people, who imagine that they have possessions, ...who now nurture an insatiable appetite for knowledge, power, health, glory, and pleasure.
>
> Only he who recognizes that he has nothing, that he cannot possess anything, that security is unattainable, who completely resigns himself and sacrifices all, who gives everything, who does not know anything, who does not want anything and does not want to know anything, who abandons and neglects all, he will receive all: the world of freedom is opened to him, the world of painless contemplation and—of nothing.[14]

Despite his praise for the life of self-abnegation, Brouwer embarked on a self-righteous campaign to reconstruct mathematical practice from the ground up so as to satisfy his philosophical convictions. Although he could easily have chosen a conventional mathematical topic, he was determined instead, to write his doctoral dissertation on the foundations of mathematics.[15] His adviser reluctantly agreed, but appalled by his prize student's insistence on injecting his strange and irrelevant ideas into his dissertation, he wrote:

> ...I have again considered whether I could accept Chapter II as it stands, but honestly Brouwer, I cannot. I find it all interwoven with some kind of pessimism and mystical attitude to life which is not mathematics, nor has anything to do with the foundations of mathematics.[16]

For Brouwer, mathematics existed in the consciousness of the mathematician, and was ultimately derived from *time* as the "mathematical Primordial Intuition." The real mathematics is in the mathematician's intuition and not in its expression in language. Far from mathematics being logic (as Frege and Russell had maintained), logic is derived from mathematics. For Brouwer, Cantor's belief that he had found different sizes of infinity was so much nonsense, and his continuum problem was a triviality. Hilbert was mistaken in claiming that consistency is all that is needed for mathematical existence. On the contrary:

> ... *to exist* [Brouwer's italics] in mathematics means: to be constructed by intuition; and the question whether a certain language is consistent, is not only unimportant in itself, it is also not a test for mathematical existence.[17]

Echoing Kronecker's call for construction as the only valid method for establishing existence in mathematics, Brouwer went further and denounced the use of a fundamental law of logic, Aristotle's *law of the excluded middle* (which simply asserts that any proposition is either true or false) when applied to infinite sets.[18] For Brouwer, some propositions can neither be said to be true or to be false; these are propositions for which no method is currently known by means of which this can be decided one way or the other. Hilbert's original proof of Gordan's conjecture used the law of the excluded middle in the way mathematicians usually do: he showed that denying the conjecture would lead to a contradiction. To Brouwer such a proof was unacceptable.

After completing his dissertation, Brouwer made a conscious decision to temporarily keep his contentious ideas under wraps and to concentrate on demonstrating his mathematical prowess. The arena he selected was the burgeoning new field of topology. He obtained a number of deep results including his important *fixed point theorem.**

In 1910, when the 29-year-old Brouwer published this fundamental principle, he had already won Hilbert's admiration. David Hilbert was greatly impressed and even invited the younger man to join the editorial board of his prized journal, the *Mathematische Annalen*, an invitation he would live to regret. After obtaining a regular academic appointment at the University of Amsterdam in 1912 (with the help of Hilbert, who was one of those who wrote on his behalf), Brouwer felt free to return to his revolutionary project which he was now calling *intuitionism*.

Hermann Weyl was Hilbert's prize student, one of the great mathematicians of his century, the one eventually chosen to take Hilbert's place at Göttingen. His interests spanned mathematics, physics, philosophy, and even art. Much to Hilbert's dismay, Weyl convinced himself that the foun-

*In 1994 the Nobel Prize in economics was awarded to two economists and the mathematician John Nash. The award to Nash was for a theorem from his doctoral dissertation of 1950 that had found numerous applications in economics and elsewhere. In this dissertation, Nash had made ingenious use of the Brouwer fixed point theorem.

dation for dealing with limit processes that had been erected by Weierstrass, Cantor, and Dedekind was shaky. He couldn't bring himself to accept the system of real numbers on which all of it was based. The entire edifice, he famously declared, "is a house built on sand."[19]

Weyl's own attempt to reconstruct the continuum of real numbers, *Das Kontinuum* ultimately failed to satisfy him, and, when he learned how Brouwer proposed to go about it, he was hooked. "... Brouwer, that is the revolution," he declared.

This was too much for Hilbert, who may well have thought, "Et tu Bruté!" The 1920s were indeed revolutionary times in Germany. The country had lost the First World War and had been forced to accept the humiliating Versailles Treaty. The Social Democratic government that took power after the abdication of the Kaiser was beset by severe economic problems and by attempts from the left and right to overthrow it. Extreme rhetoric was to be heard on all sides. In this heady atmosphere, in an address delivered in 1922, Hilbert responded to his former student's desertion as if to treason:

> What Weyl and Brouwer are doing amounts in essence to taking the path once laid out by Kronecker: they seek to provide a foundation for mathematics by pitching overboard whatever discomforts them and declaring an embargo à la Kronecker. But this would mean dismembering and mutilating our science, and, should we follow such reformers, we would run the risk of losing a large part of our most valued treasures. Weyl and Brouwer outlaw the general notion of irrational number, of function, even of number-theoretic function, Cantor's [ordinal] numbers of higher number classes, etc. The theorem that among infinitely many natural numbers there is always a least, and even the logical law of the excluded middle, e.g., in the assertion that either there are only finitely many prime numbers or there are infinitely many: these are examples of forbidden theorems and modes of inference. I believe that impotent as Kronecker was to abolish irrational numbers (Weyl and Brouwer do permit us to retain a torso), no less impotent will their efforts prove today. No! Brouwer's [program] is not as Weyl thinks, the revolution, but only a repetition of an attempted putsch with old methods, that in its day was undertaken with greater verve yet failed utterly. Especially today, when the state power is thoroughly armed and fortified by the work of Frege, Dedekind, and Cantor, these efforts are foredoomed to failure.[20]

Noting the martial flavor of Hilbert's diatribe, one might have thought that he was among those numerous Europeans who greeted the coming of war in 1914 with frenzied euphoria. But this was far from being the case. From the first Hilbert let it be known that he regarded the war as foolish. In August 1914, 93 famous German intellectuals addressed a manifesto to "the civilized world" in response to the indignation in England, France, and the United States over the actions of the German military in Belgium, asserting: "It is not true that we have criminally violated the neutrality of Belgium ... It is not true that our troops have brutally destroyed Louvain."

Hilbert had been asked to sign, but he refused, insisting that he just didn't know whether the charges were true. In 1917, five years before Hilbert's denunciation of Weyl and Brouwer, while bloody trench warfare was still in the process of devouring a generation of European men, Hilbert published a laudatory obituary notice about the recently deceased great French mathematician Gaston Darboux. When student demonstrators, gathered in front of his house, called for the repudiation of this memorial to an "enemy mathematician," Hilbert responded by demanding and receiving an official apology.[21]

When opposition arose to the proposal that the brilliant young mathematician Emmy Noether be appointed Privatdozent at Göttingen, on the grounds that this could lead to a woman becoming a professor and a member of the university senate, Hilbert declared: "I do not see that the sex of a candidate is an argument against her admission as a Privatdozent. After all, the Senate is not a bath-house."[22] In September 1917, while Germany and its neighbor France were engaged in doing their best to slaughter one another's citizens, Hilbert delivered a lecture in Zürich entitled *Axiomatic Thought* that began with the provocative sentence:

> Just as in the life of peoples, one folk can only flourish if things also go well with all of its neighbors, and as the interest of the nations require not only that order reigns within each individual nation, but also that relations among the nations be properly arranged, so is it also in the life of the sciences.[23]

Metamathematics

The problem of the consistency of arithmetic was problem number two in Hilbert's 1900 address to the International Congress of Mathematicians. But it was only during the 1920s that Hilbert formulated his serious approach to the problem. His student Wilhelm Ackermann, and his assistant Paul Bernays worked closely with him, and John von Neumann also contributed.*

*John von Neumann, one of the great mathematicians of the twentieth century, was born in Budapest in 1903. A child prodigy, he grew up in a wealthy family willing to devote resources to nurturing his talent. He worked in a great variety of topics in pure and applied mathematics (including mathematical physics and economics). He became a member of the Institute for Advanced Study in Princeton at its founding in 1933, and held this position until his death in 1957. During the Second World War he became heavily involved in military problems, including the atomic bomb project in Los Alamos. This interest, which continued into the Cold War period, led to his concern for the development of advanced computational equipment.

Hilbert began with the logical system of the Whitehead-Russell *Principia Mathematica*, and at first went along with the Frege-Russell goal of defining *number* in purely logical terms. But he was soon led to abandon this goal as untenable, while continuing to see the symbolic logic they had developed as crucial.

In Hilbert's new program, mathematics and logic were to be developed together in a purely formal symbolic language. Such a language may be thought of being viewable from the "inside" or from the "outside." From the inside, it is just mathematics, with each tiny deductive step made utterly explicit. But from the outside it is only a lot of formulas and symbol manipulation, which may be handled without regard to meaning. The task was to prove that no pair of formulas could be derived in the language that explicitly contradicted one another, or equivalently (as it turns out), that such formulas as $1 = 0$ or $0 \neq 0$ cannot be derived.

The criticisms by Poincaré and Brouwer had to be faced: nothing worthwhile could result from a consistency proof that relied on the methods it was intended to secure. Hilbert's bold idea was a brand-new kind of mathematics that he called *metamathematics* or *proof theory*. The desired consistency proof was to be carried out within metamathematics. While within the formal system the fullest unrestricted use of mathematical methods of every kind was to be permitted, metamathematical methods were to be restricted to methods beyond dispute, methods Hilbert called "finitary." Thus Hilbert hoped to be able to thumb his nose at Brouwer and Weyl, saying in effect, "I've proved that mathematicians will never run into a contradiction using their usual methods, and I've proved it using methods of which even you approve." Or as von Neumann actually put it, "Proof theory should construct, so to speak, classical mathematics on an intuitionistic basis and in this way reduce intuitionism ad absurdum."[24]

Among the mathematical "treasures" his methods would rescue, Hilbert emphatically included Cantor's transfinite numbers, of which he said, "This appears to me to be the most admirable flower of the mathematical intellect and in general one of the highest achievements of purely rational human activity."[25] Dismissing the criticisms of Brouwer and Weyl, he proclaimed, "No one shall be able to drive us from the paradise that Cantor created for us."[26]

Prepared to concede that Hilbert's program might well succeed in its own terms, Brouwer remained unimpressed: "... nothing of mathematical value will thus be gained: an incorrect theory, even if it cannot be inhibited by any contradiction that would refute it, is none the less incorrect, just as a criminal policy is none the less criminal even if it cannot be inhibited by any court that would curb it."[27]

The battle of words between Hilbert and Brouwer escalated to a war of deeds when Hilbert resorted to quasi-legal methods to dump Brouwer from the editorial board of the *Mathematische Annalen*, leading Albert Einstein to complain about "this frog and mouse" battle.[28] The controversy between Hilbert and his collaborators, on the one hand, and Brouwer and Weyl, on the other, was certainly rooted in basic philosophical questions about the nature of knowledge. Indeed, views on both sides were heavily influenced by the ideas of Immanuel Kant. However, unlike much philosophical controversy, the positions taken by Hilbert and Brouwer took the form of *programs*, leading to quite specific problems, and thus were exposed to the possibility of refutation by events.

The main problem facing Brouwer's intuitionism was to actually carry out the reconstruction of mathematics called for in his program, to convince working mathematicians that they could carry on without the classical continuum of real numbers and without the law of the excluded middle and still not risk losing some of their "most valued treasures." However, the intuitionistic mathematics that Brouwer actually produced suffered from what Weyl much later called "an almost unbearable awkwardness ... " and made few converts.[29]

Although Brouwer never recanted his views, he felt more and more isolated, and spent his last years under the spell of "totally unfounded financial worries and a paranoid fear of bankruptcy, persecution and illness." He was killed in 1966 at the age of 85, struck by a vehicle while crossing the street in front of his house.[30] Perhaps the greatest irony in the story is that intuitionism lives on after all, not, as Brouwer had intended, as the corrected practice of working mathematicians, but rather as the study of formal logical systems that are designed to incorporate elements of his ideas.[31] Some of these systems have actually formed the basis for working computer programs that carry out formal deductions.[32]

Of course, the principal problem posed by Hilbert's program was the problem with which it all began: the consistency of arithmetic. Ackermann and von Neumann worked on this problem and achieved partial results, and it was believed that it was just a matter of sharpening technique to get the full result. In 1928, Hilbert with his student Ackermann published a skinny little textbook on logic based on the lecture courses Hilbert (with Bernays's assistance) had been giving since 1917. In this book two problems were posed about the basic logic of Frege's *Begriffsschrift*, what is called nowadays *first-order logic*.

In a sense, both problems had been in the air for some time, but it was Hilbert's insight that logical systems could be viewed from the outside that led to the sharp form in which they were stated. One of these problems was to prove that first-order logic is *complete* in the sense that any formula that viewed from the outside is valid can be derived inside the system using only the rules proposed in the textbook.

The second, which became known as Hilbert's *Entscheidungsproblem*, was to provide a method that would, given a formula of first-order logic, determine in a finite number of well-defined effective steps whether or not that formula is valid. As we shall see in Chapter 7, these two problems, especially the *Entscheidungsproblem*, brought into the twentieth century, as concrete problems for mathematicians to solve, matters about which Leibniz could only dream in the seventeenth.

In this same year 1928, Hilbert addressed an International Congress of Mathematicians in Bologna. Except when international conditions made it impossible, these Congresses took place regularly at four-year intervals. Of course, there was no Congress in 1916 because of World War I. Conferences were held in 1920 and 1924, but the postwar bitterness was so great that the Germans were not invited. It was Hilbert who insisted that the German mathematicians accept the invitation to attend the 1928 Congress, against the protests of those, like Ludwig Bieberbach (later a Nazi) and Brouwer, who wanted the meeting boycotted as a protest against the Versailles Treaty.

In his address, Hilbert posed a problem concerning a formal system based on applying the rules of first-order logic (essentially Frege's rules) to a system of axioms for the natural numbers. Nowadays this system is known as Peano arithmetic (after the Italian logician Giuseppe Peano), or PA. Hilbert asked for a proof that PA is *complete*, meaning that for any proposition that can be expressed in PA, either it can be proved in PA that the proposition is true or it can be proved in PA that the proposition is false. The solution of this problem two years later by a young logician named Kurt Gödel, was not at all what Hilbert had anticipated, and indeed turned out to have devastating import for Hilbert's program.

Catastrophe

Hilbert's wife Käthe is described by his biographers as a wise and sensible person, a loyal helper to her husband, many of whose papers were handwritten by her, a mother and dispenser of wisdom about life to the young mathematicians to whom the Hilbert house seemed always to be open.

Hilbert thought himself a man of the world and used to quip that the best possible vacation is taken with a colleague's wife. He never tired of flirtations and would try to dance with the prettiest young women when the occasion allowed. His "flames" were so notorious that at a jolly birthday celebration, impromptu verses were produced about his "loves" with a different one for each letter of the alphabet. But when it came to "K" everyone was stumped. At this point Käthe remarked, "Well, you could at least think of me for once." Immediately the following verse was generated (the very free translation is mine):

Gott sei Dank	Thanks be to God
nicht so genau,	She won't have strife.
Nimmt es Käthe	"Who cares," says Käthe,
seine Frau	And she is his wife.

Their son Franz was a source of distress (in different ways) to husband and wife. His strong physical resemblance to his father only served to emphasize that it was not accompanied by any resemblance in the mental sphere. Despite efforts to pretend otherwise, it became clear that Franz was a badly disturbed young man, and it finally became necessary to institutionalize him. Hilbert's reaction to this tragedy was that he no longer had a son; his wife felt otherwise.

In 1929 a wonderful new building to house the Göttingen Mathematical Institute opened its doors. Funding had been provided by the Rockefeller Foundation and the German government, largely as the result of Richard Courant's skillful diplomacy. But the days when Göttingen could be the world center of mathematical research were almost at an end.

When Hilbert retired in 1930, Hermann Weyl accepted an offer to take over his position. That same year Hilbert was honored by being granted the title "honorary citizen" by his birth city Königsberg. He was invited to give a special address that fall in Königsberg to a meeting of scientists and physicians, and Hilbert chose an appropriately general topic: *natural science and logic.* In a wide-ranging speech, he emphasized the crucial role that mathematics plays in science and that logic plays in mathematics. With his usual optimism, he insisted that there are no unsolvable problems. He concluded with the words:[33]

Wir müssen wissen	(We must know)
Wir werden wissen	(We will know)

During the days immediately preceding Hilbert's address, a symposium on the foundations of mathematics took place in Königsberg. The speakers were Brouwer's student and disciple A. Heyting, the philosopher Rudolf Carnap, and (representing Hilbert's proof theory program) John von Neumann. At the round-table discussion that concluded the event, a shy young man named Kurt Gödel (the subject of our next chapter) made a quiet announcement that, to those who grasped its import, signaled a new era in foundational studies.

Von Neumann got the point at once, and concluded that the jig was up, that Hilbert's program could not succeed. When Hilbert learned of Gödel's announcement his initial reaction was to become angry at what may well have seemed to him a frontal attack on his "*Wir werden wissen.*" But when Bernays came to write up the achievements of Hilbert's proof theory in two massive volumes that appeared in 1934 and 1939, Gödel's work played a prominent role.[34]

Nineteen thirty-two was the year of Hilbert's 70th birthday, and it was duly celebrated in the new Mathematical Institute building. There were toasts and music, and of course, dancing, and the old man was on the dance floor for most of the dances. Nineteen thirty-two was also the year when, the depression being in full swing, the Nazis made great gains in the elections to the Reichstag. The following January, Hitler was appointed Chancellor, and the collapse of German science followed soon thereafter. Jews were not allowed to teach, and one after another, they found their way abroad.

Richard Courant, despite his service in the German army during World War I, found himself an outcast at the Mathematical Institute for which he had done so much, and ended up at New York University where he was eventually to found another mathematical institute. The Courant Institute occupies a handsome building in Greenwich Village in New York City. Hermann Weyl, though an "Aryan," found the situation in Germany intolerable and accepted a position at the new Institute for Advanced Study in Princeton joining Albert Einstein.*

Hilbert seems to have been bewildered by the new political situation, on the one hand speaking out against the regime even as it became increasingly dangerous to do so, on the other, unable to comprehend that the vaunted German legal system was unable to protect against arbitrary assaults. At a gathering, Hilbert asked Blumenthal, his first doctoral student, what courses he was teaching. Being told that he was no longer permitted to teach, the old man reacted with indignation unable to comprehend why Blumenthal did not take legal action. Blumenthal himself made his way to Holland, but when the Germans invaded in 1940, he found himself trapped. He died in 1940 in the notorious ghetto that had been established at Theresienstadt in what is now the Czech Republic.

Hilbert died in 1943 with World War II still raging. Käthe followed two years later. On Hilbert's tombstone were the words:

Wir müssen wissen
Wir werden wissen

---

*I was fortunate to hear lectures by both of these great scientists during my graduate student days in the late 1940s. Neither lecture was outstanding as an example of scientific exposition, but that was not the point. We flocked eagerly to Fuld Hall (where the Institute for Advanced Study has its headquarters) to hear these legendary figures.

Hermann Weyl, was to introduce a series of lectures by the Japanese mathematician Kodaira. What I remember best about his lecture was the pleasure with which he spoke about mathematical ideas. Weyl's lecture was rather poorly organized while Kodaira's lectures were models of clear mathematical exposition.

Einstein's lecture was occasioned by his discovery that a set of equations for a "unified field theory" could be derived from what is called a "variational principle." He totally lost track of the time as he wrote on the blackboard, and only stopped when J. Robert Oppenheimer (the director of the Institute) called the time to his attention.

CHAPTER 6

Gödel Upsets the Applecart

In the fall of 1952, shortly after my wife Virginia and I had arrived in Princeton for a two-year stay at the Institute for Advanced Study, we were driving down "Olden Lane" approaching the Institute when we found our way blocked by an odd pair walking slowly in front of our car. The taller man was quite unkempt while the other was immaculately dressed in a business suit and carried a briefcase. As I cautiously passed them, we could see that the walkers were Albert Einstein and Kurt Gödel. "Einstein and his lawyer," Virginia quipped.

It was not only in their dress that these good friends differed. After the 1952 presidential election, Einstein declared, "Gödel has gone completely crazy ... He voted for Eisenhower."[1] To the liberal Einstein, voting for a Republican was inconceivable. Their views on some fundamental philosophical issues were also far apart.

In formulating his special theory of relativity, Einstein had been influenced by the skeptical positivism of Ernst Mach with its attack on Immanuel Kant's doctrine that our notions of space and time (although objective) are independent of empirical observation.

Gödel began reading Kant as a teenager and remained very much interested in the work of the classical German philosophers (especially Leibniz), all of his life. Indeed, in an unpublished manuscript found among his papers after his death, he maintained that relativity theory, properly understood, confirms certain of Kant's views about the nature of time.[2] Echoing the complaint of Frege and Cantor about the limitations of positivism, Gödel has let it be known that it was precisely by rejecting those ideas that it became possible for him to see connections that other logicians had overlooked, making his momentous discoveries possible.[3]

After Gödel's death in 1978, a Kurt Gödel Society, devoted to research in logic and related areas in computer science, was founded in Vienna, and that is where its meetings ordinarily take place. However, in August 1993 the society met in Brno, in the Czech Republic, where Gödel had been born 87 years earlier.

KURT GÖDEL AND ALBERT EINSTEIN
(Richard Arens)

In addition to a scientific program, the meeting featured a ceremony in which Brno's civil authorities dedicated a commemorative plaque placed on Gödel's childhood home. I well remember the occasion: we stood under our umbrellas in the chill pre-autumnal drizzle, while the inevitable speeches (in Czech) were followed by several numbers played by a local band in colorful folk costume.

Kurt Gödel was born in 1906 in Brno, then still part of the Austro-Hungarian Empire. Bertrand Russell, for some reason, believed that Gödel was Jewish. In actuality his mother's family was Protestant, while his father was nominally an Old Catholic, although they were not church-goers. Kurt's schooling was entirely in German schools. Because of his meticulous habits and his apparent unwillingness to throw anything away, an unusually complete picture of his primary schooling is available. His report cards show a student who received top grades in all subjects, and his workbooks give evidence of a curriculum heavy on drilling.

At the age of eight, Kurt fell ill with rheumatic fever, but it seems that no lasting physical damage resulted in his case. On the other hand, Gödel did become a life-long hypochondriac, very likely as the result of this illness. His older brother Rudolf stated that even as a child, Kurt Gödel showed signs of mental instability.[4]

When the Austro-Hungarian Empire was dismembered after World War I, the Gödel family found itself part of the large German-speaking minority in the newly formed Czechoslovakia. German-speaking Vienna, a mere 68 miles to the south of Brno, with its fine university, soon drew Rudolf and Kurt to it. After a rigorous secondary-school education in Brno with an almost perfect academic record, Kurt moved to Vienna in the fall of 1924, sharing an apartment with Rudolf who had moved there some time before as a medical student. Although Kurt's original intention had been to study physics, the beauty of the patterns among the integers revealed in lectures he heard on the theory of numbers persuaded him that mathematics was his true calling.

The Austrian Republic, formed from the debris of the Austro-Hungarian Empire at the end of World War I, lasted a mere 20 years before it was absorbed into Nazi Germany in 1938. These were years of tumult and confusion, with the nation often teetering on the edge of civil war between "red" (that is, social democratic) Vienna and the deeply conservative countryside. It was in this turbulent atmosphere that the famed Vienna Circle flourished.

Whitehead and Russell had developed an artificial language for mathematics in which proofs of theorems could be represented by purely symbolic formal operations. The Vienna Circle was formed in 1924 by a group of philosophers and scientists continuing in the empiricist-positivist tradition of Mach and Helmholtz. It will be recalled (see the quotations at the end of Chapter 4) that Cantor and Frege had bitterly attacked these ideas. The cricle abhorred traditional metaphysics while it believed that an important goal for philosophy should be the development and study of symbolic systems like that of Whitehead-Russell that would encompass not only mathematics, but also empirical science.

When the founder, Moritz Schlipp, was assassinated in 1936 by a deranged former student, the Nazis justified the killing on the grounds of Schlipp's supposed left-wing views. Among the other important adherents of the circle were Rudolf Carnap, who had studied with Frege, and Hans Hahn who was to be Gödel's principal teacher.*

Gödel's Doctoral Dissertation

While Bertrand Russell's ideas about the foundations of mathematics had taken concrete form in the massive three-volume *Principia Mathematica*, those of his student, the brilliant and quixotic Ludwig Wittgenstein, were presented to the world in his slim 75-page *Tractatus Logico-Philosophicus*. The ideas of these two philosophers played an important role in the ongoing discussions at meetings of the Vienna Circle.

When Gödel began attending these meetings in 1926 at the invitation of his teacher Hans Hahn, he found himself out of sympathy with much of what he heard. Even so, the heady mixture of Russell's demonstration that all of ordinary mathematics can be encapsulated in a formal logical system and Wittgenstein's emphasis on the problems of speaking about language from within language must have influenced the direction of the young Gödel's research. These concerns of Wittgenstein echoed Hilbert's stance that formal logical systems could not only represent mathematical reasoning on the inside, but could also be studied from the outside using mathematical methods.

In the courses on logic that Hilbert had been giving at Göttingen, he adopted the basic rules of logical deduction proposed by Frege in his *Begriffsschrift* and incorporated by Whitehead and Russell in their *Principia Mathematica*. In his 1928 textbook on logic (written with his student Wilhelm Ackermann), Hilbert posed the question of whether there are gaps in these rules, that is, deductive inferences that ought to be correct, but for which the rules do not suffice to obtain the conclusion from the premises.

*Carnap's doctorate was from the University of Jena where he studied under Frege. He was a leading figure in the philosophy called *logical positivism*. From 1935 on, he held positions at American universities, first at the University of Chicago and then at UCLA.

Hans Hahn, Gödel's dissertation advisor, made important contributions to a number of branches of mathematics, and was also interested in philosophical questions.

His belief was that there are no such gaps, but he wanted a proof that this is the case, that the rules are *complete*.

Gödel chose this problem for his doctoral dissertation. Although he succeeded in short order in obtaining the result Hilbert wanted, there was some irony in the situation. The techniques that Gödel used were quite familiar to the logicians of the day, but, as we shall see, their hands had been tied by the influence of the Brouwer-Weyl strictures together with Hilbert's tacit acceptance of them as appropriate in metamathematical investigations.

Logical deduction proceeds from *premises* to a *conclusion*. When the symbolic logic of Frege-Russell-Hilbert is used, each premise as well as the conclusion is represented by a logical formula, which just amounts to a string of symbols.[5] Some of these symbols stand for purely logical concepts, some serve as mere punctuation, and some refer to the specific subject matter in question. Here is a sample logical inference in which the first two lines shown are the premises and the third line is the conclusion.

Anyone in love is happy.
William loves Susan.

William is happy.

Using the logical symbolism introduced in Chapter 3, we can translate this into the language of logic as follows:

$$\frac{\begin{array}{c}(\forall x)((\exists y)L(x,y) \supset H(x)) \\ L(W,S)\end{array}}{H(W)} \tag{*}$$

In this inference the logical symbols used are $\supset$ $\forall$ and $\exists$ whose meanings are recalled in the following table:

$\supset$	if ... then ...
$\forall$	every
$\exists$	some

The letters x and y serve as *variables* which stand in (like pronouns) for arbitrary individuals in the population being considered. The L, W, H and S have meanings relevant to the particular subject matter as shown below:

$$\begin{aligned} L &= \text{the relation of loving} \\ H &= \text{the property of being happy} \\ W &= \text{William} \\ S &= \text{Susan} \end{aligned}$$

So we can read the inference as follows:

For all x, if there is a y such that x loves y, then x is happy.

William loves Susan.

William is happy.

Now what it means to say that *this inference is valid* is that no matter what underlying universe of individuals we may choose, no matter what relationship between such individuals we represent by the letter L, no matter what property of such individuals we represent by the letter H, and no matter which particular individuals we choose to designate as W and S, as long as we do this in such a manner that the two premises are both true statements, then the conclusion will be true as well. To help clarify what it means for an inference to be *valid*, it may be helpful to consider another interpretation of the same symbolic inference with a very different subject matter:

Predators have sharp teeth.

Wolves prey on sheep.

Wolves have sharp teeth.

In order to see that this example is also included under the symbolic inference (*), we let the variables x and y stand for arbitrary species of mammals, and interpret the other letters according to the table:

$L =$ relation of one species preying on another

$H =$ property of having sharp teeth

$W =$ wolves

$S =$ sheep

Thus, the symbolic inference may be read:

For all x, if there is a y such that x preys on y, then x has sharp teeth.

Wolves prey on sheep.

Wolves have sharp teeth.

Hilbert asked for a proof that for every inference that is valid in the sense just explained, there is a step-by-step proof of the conclusion from the premises using the Frege-Russell-Hilbert rules. In other words, Hilbert wanted a proof that if a proposed inference has the property:

> for any interpretation of the letters in the formulas with respect to which the premises are true statements, the conclusion is true as well,

then the Frege-Russell-Hilbert rules can be used to lead from the premises to the conclusion. In Gödel's doctoral dissertation, he succeeded in providing exactly what Hilbert had requested.

Gödel's proof was explained with a directness and clarity that were to mark his later publications as well. But although this work was an impressive achievement whose great importance became clear only with the passage of time, there was little novelty in his methods, all perfectly well known to logicians at the time. This could well lead one to wonder at the inability of the powerful team of Hilbert, Ackermann, and Bernays to find its way to a proof. Indeed, Gödel commented many years later that the theorem was an "almost trivial consequence" of results in a paper by the Norwegian logician Thoralf Skolem that had appeared in 1923, six years before Gödel's dissertation (although presumably neither Gödel nor his adviser had been familiar with this paper). In a letter written in 1967, Gödel looking back at the 1920s, referred to a "blindness ... of logicians [that] is indeed surprising." But, he continued:

> I think the explanation is not hard to find. It lies in a widespread lack, at that time, of the required epistemological attitude toward metamathematics and toward non-finitary reasoning.[6]

Following the Brouwer-Weyl criticisms (discussed in the previous chapter) of "non-finitary" reasoning, and Hilbert's defining his "metamathematics" as permitting only finitary reasoning, it was at least tacitly accepted that investigations of formal logical systems from the "outside" had to be strictly limited to finitary methods, methods to which Brouwer could not object.[7] But, in fact, Gödel's completeness theorem cannot be proved without the use of non-finitary methods. Without quarreling with the aims of Hilbert's program and its methodological restrictions, Gödel explained why non-finitary methods were appropriate in this case as follows:

> ... it was not the controversy regarding the foundations of mathematics that caused the problem treated here to surface (as was the case, for example, for the problem of the consistency of mathematics); rather, even if it never been questioned that 'naive' mathematics is correct as to its content, this problem could have been meaningfully posed within this naive mathematics (unlike, for example, the problem of consistency), *which is why a restriction on the means of proof does not seem to be more pressing here than for any other mathematical problem.*[8] [italics added]

Undecidable Propositions

The second problem on Hilbert's famous 1900 list had called for a proof of the consistency of the arithmetic of real numbers. At that time, no one had any notion of what such a proof might be like, and in particular how it could avoid the trap of circularity, how it could avoid using in the proof the very methods that the proof sought to justify. As we have seen in the previous chapter, during the 1920s Hilbert introduced his program of metamathematics: axioms to be proved consistent were to be encapsulated in a formal logical system in which a proof is only an arrangement of a finite number of symbols.

Then, the proof that no contradiction could be derived in this system was to be carried out using what Hilbert called "finitary methods" that were even more restrictive than what Brouwer would have been willing to allow. When Gödel turned to these matters after completing his doctoral dissertation, Hilbert's program seemed well on the way to success.

At the International Congress in Bologna in 1928, Hilbert had spoken about the system, nowadays called Peano's arithmetic (abbreviated PA) that encapsulates the basic theory of the natural numbers $1, 2, 3, \ldots$. When Gödel began to think about Hilbert's program, Hilbert's student Ackermann and John von Neumann seemed to be advancing towards a finitary consistency proof for PA. Both had found such proofs for a limited subsystem of PA, and it was thought that progress was blocked only by technical difficulties which would be overcome in time.

Gödel may well have believed this. In any case, he set himself the problem of proving the consistency of more powerful systems *relative to* PA. There had been a number of important relative consistency proofs, so this was a natural idea. Gödel had hoped to give a finitary reduction of the consistency of powerful systems, adequate for the arithmetic of real numbers and more, to the consistency of PA. This was very much following in Hilbert's path.

Hilbert had reduced the consistency of Euclidean geometry to that of the arithmetic of real numbers, and Gödel proposed to carry the reduction one step further. Had Gödel succeeded, a proof of the consistency of PA by Hilbert's followers would have automatically provided a proof of the consistency of the arithmetic of real numbers as well, thus fulfilling the request Hilbert had made in his second problem of 1900. But it was not to be. Gödel not only failed in this endeavor, he proved that he could not have succeeded! At the end, instead of helping to secure mathematics against the Brouwer-Weyl critique as he had evidently hoped, he effectively buried Hilbert's program.

As Gödel began to think about these matters, he found himself rethinking what it meant to view a formal logical system from the *outside* as opposed to the *inside*. Russell and Whitehead had shown quite convincingly that all of ordinary mathematics can be developed *inside* such a system. Hilbert, in his metamathematics, was proposing to use mathematical methods, severely restricted to be sure, to study such systems from the *outside*. So, why can't metamathematics be developed *inside* a formal logical system?

Viewed from the outside, these systems involve relationships among strings of symbols. On the inside, these systems can express propositions about various mathematical objects including natural numbers. Moreover, it isn't difficult to think of ways that strings of symbols can be *coded* by natural numbers. Aha! *By using such codes, the outside can be brought inside.* To illustrate the use of such codes, let us look again at how the premise "Anyone in love is happy" was symbolized:

$$(\forall x)((\exists y)L(x,y) \supset H(x)) \qquad (\dagger)$$

What we have here is an arrangement or "string" of the 10 symbols:

$$, \; L \; H \; \supset \; \forall \; \exists \; x \; y \; (\;)$$

We can use a simple coding scheme in which each symbol is replaced by one of the decimal digits, for example as follows:

,	L	H	$\supset$	$\forall$	$\exists$	x	y	(	)
↓	↓	↓	↓	↓	↓	↓	↓	↓	↓
0	1	2	3	4	5	6	7	8	9

Replacing the symbols by digits in (†) as indicated, we get the code number

846988579186079328699.

Note that not only is it easy to go from the string of symbols to the number that codes it, but it is just as easy to go in the reverse direction. Of course, when there are more than ten symbols, a different encoding must be used, but this causes no difficulty. For example, if we code each symbol by a pair of decimal digits, up to 100 symbols can be accommodated. Essentially the same methods can be applied to any formal logical system, so that the various locutions of such systems (all of which are seen from the outside to be presented as strings of symbols) can be coded via natural numbers.[9]

Gödel had no problem seeing how codes could indeed be used to develop the metamathematics of formal logical systems *inside* those very same systems. But in the process, he found himself thinking thoughts that were strictly forbidden according to precepts being promulgated in the Vienna Circle. Gödel found that there are propositions that viewed from the *outside* of such systems could be seen to be true, yet could not be proved *inside* of them.

For many adherents of the Vienna Circle, any notion of mathematical truth other than provability was meaningless, a chimera of idealistic metaphysics. Being unencumbered by such beliefs, Gödel was led to the remarkable conclusion that on the contrary, not only is there a meaningful notion of mathematical truth, but its extent goes beyond what can be proved in any given formal system.

This conclusion applied to a wide range of formal logical systems: it applied to comparatively weak systems like PA and even to systems like Whitehead and Russell's *Principia Mathematica* (abbreviated PM) that encapsulated the full power of classical mathematics. For any of these systems, there are true propositions expressible in the system, but not provable in the system. As its title indicates, in Gödel's remarkable paper, *On Formally Undecidable Propositions of Principia Mathematica and Related Systems,* published in 1931, he chose to present his results for PM, thus showing that even powerful logical systems could not hope to encompass the full scope of mathematical truth.[10]

The crucial step in Gödel's proof was his demonstration that *the property of a natural number of being the code of a proposition provable in* PM *is itself expressible in* PM. Using this fact, Gödel could construct propositions in PM that to one who knew the specific code being used could be seen to express the assertion that some proposition is not provable in PM. That is, he was able to construct propositions A that, read via the encoding, assert that some proposition B is not provable in PM.

Now, someone not privy to the code looking at A would see a string of symbols expressing some complicated and mysterious proposition about natural numbers. But via the code, the mystery vanishes: A expresses the proposition that some string of symbols B represents a proposition not provable in PM. Ordinarily A and B would be different propositions. Gödel asked the question: Could they be the same? Indeed they could, and Gödel was able to demonstrate this by making use of a mathematical trick he had learned from Georg Cantor: *the diagonal method.* By using this trick, matters could be so arranged that the proposition asserted to be unprovable and the proposition making that assertion were one and the same. In other words, Gödel had seen how to obtain a most remarkable proposition, we'll call U, with the properties:

- U says that some particular proposition is not provable in PM.
- That particular proposition is none other than U itself.
- **Therefore, U says: "U is not provable in PM."**

In the Vienna Circle, it was generally believed that the only notion of "truth" that makes sense for propositions expressed in a system like PM is that of provability according to the rules of the system. The properties of this proposition U make this belief untenable. If we are willing to assume that PM doesn't lie, that whatever is proved in PM is actually true,[11] then we can see that U is true, but not provable in PM as follows:

1. *U is true.* Suppose that it were false. Then, what it says would be false. So it couldn't be unprovable and would have to be provable, and therefore true. This contradicts the supposition that *U* was false. Hence it must be true.

2. *U is not provable in* PM. Since it is true, what it says must be true, and so it is not provable in PM.

3. *The negation of U, written ¬U, is not provable in* PM. Because *U* is true, ¬*U* must be false, and therefore also not provable in PM.

To emphasize that *U* has the property that neither it nor its negation is provable in PM, it is called an *undecidable proposition.* But it cannot be emphasized too strongly that this undecidability is only with respect to provability *inside* the system. From our *outside* viewpoint, it is clear that *U* is true.*

Now here is a puzzle: we know that *U* is *true* although unprovable in PM. Since all of ordinary mathematics is encapsulated in PM, why can't the proof that *U* is true be carried on *inside* PM? Gödel came to realize that this is almost possible, but that there is a catch. What can be proved inside PM is:

If PM *is consistent then U.*

So it is only the additional assumption that PM is consistent that blocks the proof of *U* inside PM. Since we know that *U* can't be proved inside PM, we must conclude that the consistency of PM cannot be proved in PM. However, the thrust of Hilbert's program was to prove that systems like PM are consistent using "finitary" methods that were thought to constitute a very modest subset of those available in PM. Yet Gödel had proved that even the full power of PM is insufficient to prove its own consistency. So, at least as originally imagined, Hilbert's program was dead![12]

Kurt Gödel, Computer Programmer

In the year 1930, the realization of an actual physical device that could function as a general-purpose information-processing programmable computer was still decades in the future. Yet someone knowledgeable about modern programming languages today looking at Gödel's paper on undecidability written that year will see a sequence of 45 numbered formulas that looks very much like a computer program.

The resemblance is no accident. In demonstrating that the property of being the code of a proof in PM is expressible *inside* PM, Gödel had to deal

*For an explanation of how Gödel was able to construct a statement such as *U* that asserts its own unprovability, see the appendix at the end of this chapter.

with many of the same issues that those designing programming languages and those writing programs in those languages would be facing. At the most fundamental level, contemporary computers can perform only simple basic operations on short strings of 0s and 1s.

Designers of so-called high-level programming languages face the task of providing programmers with locutions that encapsulate the highly complex operations with which they would like to work. Programs written using these locutions to be carried out by a computer must be translated into machine language—into a detailed listing of the basic operations needed to execute them. This is done by special translation programs called *interpreters* or *compilers*.*

The keystone in Gödel's proof of the existence of undecidable propositions is the fact that provability in PM can be expressed in PM itself. Gödel knew very well that he would be presenting his revolutionary results to a highly skeptical audience, and he wanted to eliminate any doubt. Thus he faced the problem of breaking down complex operations on the codes of the strings of symbols corresponding to the axioms and rules of inference of PM viewed from the *outside*, and transforming them into expressions written in the symbolic language of PM.

To solve this problem, Gödel created what amounted to a special language in which the operations needed could be developed in a step-by-step fashion.[13] Each step consisted of a definition of an operation on numbers that, via the code Gödel was using, corresponded to a parallel operation on expressions of PM. The definitions were expressed in Gödel's special language in terms of items that had already been defined in previous steps. The special language was so designed that operations introduced by such a definition were guaranteed to be appropriately expressible inside PM.

Leibniz had certainly proposed the development of a precise artificial language in which much human thought would be reduced to calculation. Frege, in his *Begriffsschrift*, had shown how the usual logical reasoning by mathematicians could indeed be captured. Whitehead and Russell had succeeded in developing actual mathematics in an artificial language of logic. Hilbert had proposed the metamathematical study of such languages. But before Gödel no one had shown how these metamathematical concepts could be embedded in the languages themselves.[14]

In addition to constructing an undecidable proposition U, Gödel wished to demonstrate that this proposition required for its statement no exotic mathematical concepts. For this purpose, Gödel used a theorem from the

---

*An *interpreter* works by translating the individual steps of a program into machine language, and actually executing each step before proceeding to the next. A *compiler* translates an entire program into machine language. The machine language program thus produced can be run as a stand-alone item, without further need for the compiler. Much contemporary commercial software is generated by compilers.

elementary theory of numbers known as the Chinese remainder to show how all the operations expressible in his special language could also be expressed in the basic language of the arithmetic of natural numbers.[15] From this it followed that the undecidable proposition U could be expressed in this basic language. What this meant specifically is that U could be written using a vocabulary that permitted only variables whose values could be any natural number, the arithmetic operations $+$ and $\times$, the symbol $=$ of equality, and the basic operations of Frege's logic, nowadays written: $\neg \supset \wedge \vee \exists \forall$. The remarkable conclusion was that even restricted to this meager vocabulary, propositions that are undecidable in PM could be constructed.

Conference at Königsberg

On August 26, 1930, at the Reichsrat Cafe in Vienna, the 24-year-old Kurt Gödel was talking to Rudolf Carnap about the *Conference on the Epistemology of the Exact Sciences* planned for Königsberg 10 days later. Carnap, almost 40, and a leading figure in the Vienna Circle, was scheduled to deliver a major address on the "logicist" program for the foundations of mathematics, a program that had reached its fullest realization in the *Principia Mathematica* of Whitehead and Russell.

Carnap's notes reveal that Gödel had told him about his sensational discovery that there were propositions about the natural numbers undecidable in *Principia Mathematica.* The two logicians (together with other participants in the *Conference*) traveled to Königsberg.

On the first day of the conference, there were three hour-long addresses on the foundations of mathematics. Carnap led with his address on logicism, and remarkably, made no mention of Gödel's new results. Carnap was followed by A. Heyting, a student of L. E. J. Brouwer, who spoke on Brouwer's intuitionism. The final address of the day was by John von Neumann, whose topic was Hilbert's program.[16]

On the second day, in addition to three more hour-long talks there were three 20-minute presentations including one by Gödel on his doctoral dissertation on the completeness of Frege's rules. Gödel's bombshell came on the third day during a round-table discussion of the foundations of mathematics. He began with a rather long but tentative discussion of what would be gained from a consistency proof for a system like *Principia Mathematica.* He asserted that even if such a system is known to be consistent, it was still perfectly possible that one could prove in the system a proposition about the natural numbers, that, viewed from outside, could be seen to be false. So mere consistency of a formal system provided no guarantee that what was proved in that system was correct.

Apparently, a favorable comment by von Neumann encouraged him to go further. Gödel went on to assert that assuming the consistency of systems like *Principia Mathematica*, "one can even give examples of propositions" of a simple arithmetic form that are true, but unprovable in such a system. "Therefore," he continued, "if one adjoins the negation of such a proposition to" *Principia Mathematica*, one obtains a consistent system in which a false proposition is provable.[17]

John von Neumann seems to have grasped immediately the import of what Gödel had done and indeed sought him out for discussion at the session's end. There is no evidence that anyone else realized what had happened. Von Neumann continued to think about the matter and convinced himself that (for reasons explained above) it follows from Gödel's result that consistency is unprovable, and concluded that that was the end of Hilbert's program.

By the time a letter from von Neumann arrived with this information, Gödel had already submitted for publication his own paper containing this same conclusion. Von Neumann's letter thanking Gödel for a preprint of this paper said, perhaps ruefully, "Since you have established the unprovability of consistency as a natural continuation and deepening of your earlier results, of course I will not publish on that subject."[18] Logic and the foundations of mathematics had been one of von Neumann's important interests. He became a good friend of Gödel, lectured widely on Gödel's work, and spoke of him as the greatest logician since Aristotle.[19]

Von Neumann stopped working in logic. When his interests returned to logic over a decade later, it was in logic as embodied in hardware: the all-purpose digital computer.

One of von Neumann's collaborators in his later work with computers, reports the following amusing story that von Neumann used to tell about his efforts to prove the consistency of arithmetic:

> At the end of a day's work [von Neumann] would go to bed and very often awaken in the night with new insights.... In this case he was busily engaged in trying to develop a proof [of the consistency of arithmetic] and was unsuccessful! One night he dreamed how to overcome his difficulty and carried his proof much further along ... The next morning he returned to the attack, again without success, and again that night retired to bed and dreamed. This time he saw his way through the difficulty, but when he arose ... he saw there was still a gap ...

Things would have worked out differently, von Neumann quipped, if he had dreamed a third night![19]

The conference at which Gödel dropped his bombshell was but an adjunct to the major event taking place in Königsberg that week: a convention of the Society of German Scientists and Physicians. The opening address was delivered by David Hilbert the day following the round table. This was the occasion in which Hilbert articulated the slogan still on his tombstone in which he declared his faith that all mathematical questions must and will be answered: "*wir müssen wissen; wir werden wissen,*" [we must know; we shall know].

Gödel's incompleteness theorem shows that if mathematics is restricted to what can be encapsulated in specific formal systems like PM, then Hilbert's faith was in vain. For any specific given formalism there are mathematical questions that will transcend it. On the other hand, in principle, each such question leads to a more powerful system which enables the resolution of that question. One envisions hierarchies of ever more powerful systems each making it possible to decide questions left undecidable by weaker systems.

Although all of this is incontrovertible as a matter of theory, it is less clear to what extent it will ever become a matter of mathematical practice. Gödel has left as his legacy to mathematicians the task of learning to use these more powerful systems in settling intractable problems. Although some courageous researchers have been working along these lines, most mathematicians remain unaware of these issues, and some experts greet this work with extreme skepticism.[20]

Love and Hate

One of Gödel's fellow students in Vienna, Olga Taussky-Todd, who later became a prominent number theorist, reports that Gödel's ability was well-recognized among the students, and that when a student experienced difficulty, he always stood ready to help. She tells the following amusing anecdote:

> There is no doubt that Gödel had a liking for members of the opposite sex, and he made no secret about this fact. ... I was working in the small seminar room outside the library in the mathematical seminar. The door opened and a very small, very young girl entered. She was good looking ... and wore a beautiful, quite unusual summer dress. Not much later Kurt entered and she got up and the two of them left together. It seemed a clear show off on the part of Kurt.[21]

Gödel met Adele, the woman who was to be his life's companion, during his student days, a decade before they married. At the time, she was still married to her first husband and worked as a dancer.* His parents could hardly have been pleased by his choice. It was not only that she was six

*According to one account she danced at *Der Nachfalter*, a nightclub whose name "the moth" was intended to suggest shadowy creatures of the night. Another version has it that she was a ballet dancer.

years older than Kurt and a Roman Catholic as well. It appears that female dancers in Vienna had the reputation, deserved or not, of being sexually available for a modest sum.[22]

Perhaps for these reasons, Kurt was quite circumspect about his relationship with Adele which seems to have been a close intimate one for some time before their marriage, and when they finally did marry, her existence came as quite a surprise to Gödel's colleagues.[23] Rudolf (who himself had remained a bachelor) wrote shortly after his brother's death, "I would not presume to pass judgment on my brother's marriage."[24]

Marital happiness remains a great mystery, and the prognostications of those older and presumably wiser are often wrong. This was the case with the Gödels whose marriage proved to be long-lasting and happy.

Gödel's attempts to develop a professional career in Austria occurred against the background of tumultuous and calamitous political, social, and economic events. The German-speaking state that emerged from the debris of the Austro-Hungarian Empire at the close of World War I was forbidden by the Allies to do what most Austrians desired: to unify with Germany. At any rate, independent democratic Austria did not last very long. A low-intensity civil war between the fascist *Heimwehr* and the social democratic *Schutzbund* reached a climax in 1927. When an old man and a child were killed by reactionaries and a jury refused to convict the killers, a mass demonstration called by the Social Democrats led to the burning to the ground of the Ministry of Justice building and the deaths of almost 100 people.

By the end of 1929, the president of the republic had obtained the power to rule by emergency decree. Meanwhile the great world-wide economic crisis (in the U.S. known as the Great Depression) was making moderation seem irrelevant. The Dollfuss regime, elected in 1932, took an authoritarian turn ending any meaningful role for parliament.

Things went from very bad to immeasurably worse. In early 1934, with Hitler already in power in Germany, all political parties were abolished except for Dollfuss's Fatherland Front. A few months later Dollfuss was murdered by Austrian Nazis attempting unsuccessfully to seize power. His successor, Schuschnigg, kept Hitler at bay for a few years with the help of Mussolini. But the end came in March 1938 when Austria was absorbed into Nazi Germany.

Gödel began the long academic climb with an official appointment as Dozent in February 1933. In the meantime he had been quite active in the logic seminar run by his thesis advisor Hahn as well as in an ongoing colloquium run by the mathematician Karl Menger (who was also active in the Vienna Circle).

A considerable number of Gödel's interesting results from this period, some of them quite important, were published as brief articles in the proceedings of Menger's colloquium.[25] Gödel's first course as a Dozent was given during the summer of 1933 under difficult conditions. The university had to be closed one day because of Nazi activities, and there was a week when Nazi terrorist bombs exploded in various parts of Vienna.

When an offer came to spend the academic year 1933–34 at the newly formed Institute for Advanced Study in Princeton, Gödel could hardly have turned it down. Not only would he escape the madness at home, but also he could look forward to being with such stellar colleagues as Albert Einstein and John von Neumann. However, the prospect of leaving family and friends (and perhaps especially Adele) for most of a year, surely provoked some anxiety in the shy hypochondriacal young man. Indeed, after setting out to meet the ship scheduled to take him across the Atlantic, he decided that he had a fever and turned back. Only family persuasion got him to catch another ocean liner and make the voyage.

Mathematical Drama in Princeton in the 1930s

In the 1930s (and indeed through the 1950s) the Princeton mathematics department was housed in Fine Hall, a low level attractive red-brick building.* At the time, Fine Hall housed not only the mathematics faculty of Princeton University, but also the mathematicians who were part of the recently established Institute for Advanced Study. The great influx to the United States of scientists fleeing the Nazi regime had begun. The concentration of mathematical talent at Princeton during the 1930s came to rival and then surpass that at Göttingen. Among those seen in the corridors of Fine Hall were Hermann Weyl, Albert Einstein, and John von Neumann, whose interests had moved very far from von Neumann's work on Hilbert's program for the foundations of mathematics.

The logician Alonzo Church was also present as a member of the Princeton University faculty. Stephen Kleene and Barkley Rosser, who were to have distinguished careers as mathematicians, were there as Church's doctoral students.†

Mathematicians have been dealing with algorithms for numerical calculations since ancient time. We all learned algorithms for adding and multiplying numbers and working with fractions during our childhood. But when

*The building where Princeton's mathematics department is housed today is also called Fine Hall; it is visible as a concrete tower from U.S. Highway 1, a mile away.

†Alonzo Church (1903–1995) played a crucial role in the development of a flourishing research effort in logic in the United States. He established the influential *Journal of Symbolic Logic* and served as its editor for over 40 years. Among Church's 31 doctoral students were Alan Turing and, incidentally, me as well.

Church introduced a new notation in connection with a familiar mathematical concept, he surely had no idea that he was embarking on a path that was to lead to a characterization of what is algorithmically possible, and that he would be able to exhibit an example that he would actually prove to be algorithmically unsolvable. This new notation concerned mathematical *functions.*

Typically a function is presented as a mathematical formula containing one or more letters. When these letters are each replaced by a number, a value can be calculated, the *value* of the function. But the function itself is not the formula, but is rather the association the formula provides between a given value and the calculated result. As a simple example, we can consider the expression $x^2 + 3$. If we let f stand for the function specified by this formula, we could write

$$f(x) = x^2 + 3.$$

Then, $f(1) = 4,\ f(3) = 12$ etc. Church decided that a notation was needed to show the function as arising from the expression. Using the Greek letter λ (lambda), he would write:

$$f = \lambda x[x^2 + 3].$$

Church designed a complete formal system for the foundations of mathematics incorporating his λ notation. He would have seen it as much simpler than the rather ponderous *Principia Mathematica* of Whitehead and Russell. And he thought it would succeed in avoiding the paradoxes like the one Russell had brought to Frege's attention. In this he was mistaken: his students Kleene and Rosser proved that their teacher's system was inconsistent, that it led to contradictions.

One could imagine him echoing Shakespeare's Julius Caesar's "*Et tu Brute*" when Caesar recognized his protegé among the conspirators stabbing him. Unlike Frege, Church did not give up when an inconsistency was found in his system. He thought, quite correctly, that the λ notation was worth salvaging from the wreckage. He developed a much less ambitious system, the λ-*calculus*, and was able to prove that that system is consistent.

There was no possibility that this bare-bones system could provide a foundation for all of mathematics, but one could ask whether any significant mathematics at all could be developed within Church's λ-calculus. As a language it was built from letters serving as variables, punctuation marks, and, of course λ. In particular there was no obvious way to represent numbers, even the natural numbers $1, 2, 3, \ldots$. Now of course, the natural numbers can be represented in different ways. The Arabic numerals

$$1, 2, 3, 4, 5, 6, 7, 8, 9, 10, 11, \ldots, 99, 100, \ldots$$

in our familiar decimal notation provide one such representation. The Roman numerals

$$\text{I, II, III, IV, V, VI, VII, VIII, IX, X, XI}, \ldots, \text{IC, C} \ldots$$

provide another. Algorithms for working with numbers will depend very much on how they are represented. It's much easier to multiply two numbers represented by Arabic numerals than as Roman numerals! But any systematic infinite sequence of strings of symbols could be used. Faced with the meager resources of the λ-calculus, Church and Kleene arrived at the following way to represent the natural numbers:

$$\begin{array}{ccl} 1 & \rightarrow & \lambda x[\lambda y[x(y)]] \\ 2 & \rightarrow & \lambda x[\lambda y[x(x(y))]] \\ 3 & \rightarrow & \lambda x[\lambda y[x(x(x(y)))]] \\ 4 & \rightarrow & \lambda x[\lambda y[x(x(x(x(y)))])] \\ . & \rightarrow & \ldots \\ . & \rightarrow & \ldots \\ . & \rightarrow & \ldots \end{array}$$

These formulas look complicated, but the pattern should be clear. So to calculate $2 + 3$ in the λ-calculus, beginning with

$$\lambda x[\lambda y[x(x(y))]] \text{ and } \lambda x[\lambda y[x(x(x(y)))]]$$

one would somehow need to get the result

$$\lambda x[\lambda y[x(x(x(x(x(y)))))]].$$

But it wasn't at all obvious that algorithms could be developed within the λ-calculus to carry out addition, multiplication and other arithmetic operations. Kleene undertook the investigation of arithmetic in the λ-calculus as the topic of his doctoral dissertation. Subtraction seemed especially difficult. Many years later Kleene laughed as he explained that the idea for a λ-calculus algorithm for subtracting 1 occurred to him when he was sitting in his dentist's chair.

Church and Kleene decided to call a function λ-*definable* if there is a λ-calculus algorithm for calculating its values. After such initial difficulties, Kleene became so good at developing λ-calculus algorithms that he could prove that just about any function on the natural numbers he could think of was λ-definable. And Church began to wonder how far this could go.[26]

Having finally overcome his reluctance to leave Vienna and his bride Adele, Gödel arrived in the U.S. in the fall of 1933. During the spring of 1934 he was at Princeton where he gave a series of lectures on undecidability. Kleene and Rosser took excellent notes on these lectures, and presumably Church attended as well.

In connection with metamathematical concepts (like being the code number of a proof), Gödel had introduced a class of functions defined on the natural numbers that he had called *recursive*. He chose this name because functions belonging to this class were typically defined by specifying their value for an initial input value, and then specifying how, knowing the value of the function for a given input value, to specify the value of the function for the next input value. He remarked in these lectures that the recursive functions had the important property that their values could be computed by a "finite procedure," or as we would say, by an algorithm.

He went further and suggested that the class of recursive functions could be extended to a larger class, still embodying the idea of using recursion that would include *all* functions defined on the natural numbers whose values could be calculated by an algorithm. And, as a step in that direction, he defined a class of functions he called "general recursive."

At that time Church had a conversation with Gödel in which in which they talked about what Church called "effectively calculable," or as we would say, calculable by an algorithm. Gödel raised the question of whether all effectively calculable functions are *general recursive* in the sense he had defined in his lectures. Relying on the work on λ-definability, Church had previously proposed to Kleene that λ definability was an appropriate definition of effective calculability. He now made the same proposal to Gödel. At this point Gödel was not yet ready to accept either general recursiveness functions or λ-definability as equivalents of effective calculability. Kleene set to work to find more evidence.

Studying Gödel's general recursive functions. he found that they were related in a very simple way to functions in Gödel's original narrower class of recursive functions which Kleene proposed to now call the *primitive recursive functions.*[27] He also proved that the the λ-definable functions and the general recursive functions, although defined so differently, were **exactly the same functions.**

Basing himself on all of this, Church published a paper in which he declared that these functions are exactly those that are effectively calculable, an assertion that became known as *Church's Thesis.* Church stressed that the remarkable fact that two such different conceptions turned out to yield the very same class of functions furnished evidence for this thesis. Relying on his thesis, Church was able to exhibit a specific problem which he declared to be *unsolvable* in the sense that that there was no λ-definable procedure (or equivalently no general recursive procedure) for solving it. He gave his paper the provocative title *An Unsolvable Problem of Elementary Number Theory.*[28] Meanwhile, on the other side of the Atlantic, the young Alan Turing quite independently working on the same issues. When his machine-oriented approach (to be discussed in the following chapter) turned out to be equivalent to general recursiveness and λ-definability, Church had even more evidence that his thesis was correct.

Back in Vienna

Afew months after Gödel's return to Vienna in June 1934, he suffered a "nervous breakdown" and spent some time in the Purkersdorf Sanitarium "an establishment for the well-to-do, part spa, part clinic, part rest home" where he was examined by the Nobel-prize-winning psychiatrist Julius Wagner-Jauregg.[29]

Gödel's return was to an Austria assailed by dismal events. The Nazi attempted takeover and Dollfuss's assassination occurred late in July, one day after Hans Hahn, Gödel's dissertation supervisor, died of complications of cancer surgery. At the university, things were deteriorating. Administrators were required to join the fascist Fatherland Front, and there was widespread firing of professors who were thought to be on the left and even firings of some apolitical Jewish scholars. There is no way to know what role these events may have played in Gödel's breakdown.

With the advantage of hindsight it is easy to see the menace in the steady advance of fascism. But to those who would have chosen to flee had they been gifted with knowledge of the future, matters were not so simple. One could always hope that things would work out. Gödel's brother noted that none of his family members was "very interested in politics," and so, they didn't understand the significance of Hitler coming to power in Germany in 1933. However, he went on:

> Two events quickly opened our eyes: the murder of Chancellor Dollfuss and the murder (by a National Socialist student) of the philosopher Professor Schlick in whose circle my brother had moved.[30]

While remaining in touch with the Institute in Princeton about future possibilities, Gödel continued to pursue an academic career in Vienna. He gave his second course at the university beginning in May 1935, and in September of that year set out once again for a visiting appointment in Princeton. This time he did not remain in America very long. Prostrated by a deep depression, he resigned his appointment and returned home early in December.

Gödel later spoke of 1936, the year of Schlick's assassination, as the worst year in his life. His mental condition continued to be poor, and he spent much time in sanatoria. But 1937 marked a great improvement. In June, while giving a course at the university in set theory, Gödel achieved a momentous breakthrough in his work on Cantor's continuum hypothesis, the first problem on Hilbert's famous 1900 list. (More about that later.)

Hitler's invasion of Austria and its absorption into Germany took place in March of 1938, and Gödel set out for his third visit to America in October leaving behind Adele, his bride of just over two weeks.* This time his year in America was quite fruitful. After spending the fall semester in Princeton, where he lectured on his discoveries regarding Cantor's continuum hypothesis, he took up a visiting appointment for the spring term at Notre Dame University where his old colleague Karl Menger had settled after fleeing Vienna. But when the academic year was over, he returned to Vienna and to Adele late in June 1939, a little over two months before the German invasion of Poland that was to precipitate World War II.

Gödel returned to a Vienna, now an integral part of Nazi Germany, being systematically remade as part of Hitler's "New Order." At the university, the position of Dozent had been abolished and a new position called "Dozent neuer Ordnung" (that is, Dozent of the New Order) had been put in its place. The new position did carry a small salary, but it required a new application, and the candidate had to pass muster with respect to political views and racial purity. In September, shortly after war broke out, Gödel did apply. To his surprise and indignation, his application was not approved. The report to the dean from the bureaucrat in charge of dozent applications noted that Gödel had worked under "the Jewish Professor Hahn" and that he had moved in "Jewish-liberal" circles. On the other hand he had never been known to say anything "against National Socialism." Under these circumstances it was impossible to approve the application or turn it down. An undecidable proposition!

Another serious blow came when, after months of delay, Gödel was called up for a physical to determine his fitness for military service. Once again he was surprised: he was pronounced fit for "garrison duty." Somehow amidst all of this, in November, he and Adele moved from their suburban rental flat into a recently purchased apartment in the city.[31] Gödel's apparent obliviousness to what was going on around him can only be described as pathological denial.

This is illustrated by the tale recounted by Gustav Bergmann, a member of the Vienna Circle and a Jew, one of a stream of Jewish refugees arriving in America. Shortly after his landing in October 1938, he was invited to lunch by Gödel (then visiting in Princeton) and was astounded to hear Gödel inquire, "And what brings you to America, Herr Bergmann?"[32] It seems that what finally brought Gödel's precarious situation home to him,

*There is some reason to believe that there was some plan afoot for the newlyweds to travel together to Princeton. See Dawson (1997, pp. 128–129).

shortly after his move, was his being set upon in the street by a bunch of rowdies who struck him and knocked off his glasses.[33]

After Germany's rapid conquest of Poland, the winter of 1939–1940 became known as the period of "phony war." The German onslaught on Western Europe that resulted in the defeat of France was still months away. The attack on Russia was not to take place until June 1941. In fact, Germany had signed a non-aggression pact with the Soviet Union, and Stalin's Russia was supplying Germany with products useful to the military.

It was in December 1939 that Gödel finally decided to make an all-out effort to leave Europe. In order to do this, he needed to obtain exit permits for Adele and himself from the German authorities and a visa from the U.S. authorities. Neither was easy. The newly appointed director of the Institute for Advanced Study in Princeton, Frank Aydelotte, was the hero of this endeavor. In approaching the U.S. State Department he was not above stretching the truth. In his correspondence, he wrote "Professor Gödel" although he knew perfectly well that Gödel was no professor. In answer to a question about what Gödel's teaching duties at the Institute would be, Aydelotte calmly lied, saying "Professor Gödel's responsibilities" would "involve teaching" but at an advanced, hence informal level.

In addition, Aydelotte wrote the German Embassy in Washington emphasizing that Gödel was an "Aryan" and one of the greatest mathematicians in the world. This did the trick: all necessary documents were forthcoming, and the Gödels could leave. However, the Atlantic crossing being deemed too dangerous, they traveled the long way around: through Siberia to Japan, then across the Pacific, and finally by train to Princeton, arriving in mid-March.[34]

One of the first to greet Gödel was Oskar Morgenstern, who was to become one of his best friends. Morgenstern, an economist, had known Gödel casually in the Vienna Circle and had accepted a position as professor at Princeton University after he was fired from his leading position in Austria. Eagerly inquiring about the current situation in Vienna, he was taken aback when Gödel replied, "The coffee is wretched."[35]

Hilbert's Dictum

At the head of Hilbert's problem list in his 1900 address was Cantor's continuum hypothesis. This is the assertion that infinite sets of real numbers come in just two sizes: *small* and *large*. The "small" infinite sets of real numbers are those that are just as large as the set of natural numbers, meaning that such sets can be matched up in a one-to-one fashion with the set $\{1, 2, 3, \ldots\}$. The "large" sets are those that can be matched up in a one-to-one fashion with the set of all real numbers. The continuum hypothesis is the statement that every infinite set of real numbers must be of one or the other of these types, so that there are none whose size is in between. (In the language of Cantor's transfinite cardinal numbers, the assertion is: *the cardinal number of every infinite set of real numbers is either* $\aleph_0$ *or* $\mathcal{C}$.)

In his address, Hilbert said that the continuum hypothesis was "very plausible" but that "in spite of the most strenuous efforts, no one has succeeded in proving" it.[36] Hilbert returned to the problem a quarter of a century later, claiming that he could use his metamathematics to prove the continuum hypothesis. However, this turned out to be an illusion. In 1934 a treatise by the Polish mathematician Waclaw Sierpinski was entirely devoted to propositions that had been found to be equivalent to the continuum hypothesis, or to be related to it in other ways. Yet, despite all of these continuing "strenuous efforts," it remained undecided whether the continuum hypothesis was true or false.

Gödel came to believe that the continuum hypothesis was undecidable from the available formal systems serving as foundations for mathematics. Such systems included not only Russell and Whitehead's PM, but also systems based on axioms for set theory. Gödel was able to justify his belief only in part: in 1937, he saw how to prove that in these systems it is not possible to *disprove* the continuum hypothesis.[37] Although he was convinced that it would turn out that it would be equally impossible to *prove* the continuum hypothesis in these systems, he never was able to prove that this is actually the case. (Gödel was vindicated a quarter of a century later when Paul Cohen developed powerful new methods by means of which he was able to show that the continuum hypothesis is indeed undecidable from the systems in question.)

In his address in Paris in 1900 and again in the retirement address he gave in Königsberg in 1930, Hilbert had proclaimed his faith in the solvability of every mathematical question. Was the continuing inability of mathematicians to resolve Cantor's continuum problem an indication that Hilbert had been wrong? The undecidable propositions Gödel had found involving natural numbers were undecidable *inside* the formal systems in question, but as we have seen, viewed from the outside, they were clearly true.

But the continuum hypothesis was different. Gödel's work provided no hint regarding the truth or falsity of the continuum hypothesis. Up to this point Gödel had been unencumbered by narrow foundational views, able to plough ahead using whatever mathematical methods were needed. But now his results forced him to think about the philosophical implications of what he had done.

The specific individual real numbers with which mathematicians ordinarily deal, such as π and $\sqrt{2}$, can be defined in formal systems like PM.

But, as was already clear in Cantor's time, the cardinal number of the set of all possible definitions in such systems is only $\aleph_0$ while the cardinal number of the set of all real numbers is $\mathcal{C}$, which as we know, is larger. So most real numbers have no definition: they are undefinable.

This is spooky. How can you count things that you can't define? Does it make sense to talk about sets of real numbers when some of the numbers in such a set are undefinable? Maybe the undecidability of the continuum hypothesis (conjectured by Gödel and later proved by Paul Cohen) is telling us that the continuum hypothesis does not have a clear meaning, that it is inherently vague. Dealing with this issue is to face starkly the question of the role of the actual infinite in mathematics, the very issue that Frege had predicted would lead to a "momentous and decisive battle."[38]

Manuscripts for lectures on Gödel's work on the continuum hypothesis that he gave shortly after obtaining his results on the problem show that he was equivocal. The continuum hypothesis, he suggested, might well turn out to be "absolutely undecidable," showing that Hilbert had been mistaken in believing that every mathematical problem could be solved. In the early 1940s, Gödel moved on to philosophical studies, in part no doubt to help him come to terms with his views about infinite sets. He became especially devoted to Leibniz, the classical philosopher with whom he felt the greatest affinity.

Members of the Institute for Advanced Study were under no obligation to give lectures, work with students, or even publish. Gödel responded to this relaxed atmosphere by lecturing or publishing only in response to very specific invitations. An important source of such invitations was the *Library of Living Philosophers*, a series of books each devoted to a living philosopher. Each volume was a collection of invited essays about the ideas of the philosopher in question, followed by rejoinders by the philosopher himself.

Gödel was invited to contribute to three of these volumes: the ones on Bertrand Russell, Albert Einstein, and Rudolf Carnap. The Russell volume appeared in 1944 with Gödel's contribution, a rather shocking essay. After an incisive discussion of Russell's mathematical logic, Gödel announced that sets and concepts may be "conceived as real objects ... existing independently of our definitions and constructions. ... the assumption of such objects is quite as legitimate as the assumption of physical objects and there is quite as much reason to believe in their existence." So much for vagueness!

Three years later, in an invited expository article on the continuum hypothesis, Gödel reiterated his belief in the genuine existence of sets, emphasized that existing foundational systems were necessarily incomplete and capable of extension, and predicted that new axioms would be found that would finally and definitively settle the continuum hypothesis, by enabling one to prove that *it is false*.[39]

Until his work on the continuum hypothesis, Gödel's interactions with philosophical issues had consisted mainly of ignoring the scruples of others that were preventing them from seeing what was clear to him. But now he was treading deep philosophical waters. What are numbers anyway? Are they a mere human construct or do they have some kind of objective existence? Was $2 + 2 = 4$ true before there were people on the planet to assert it? These issues have been debated for centuries. The doctrine that abstract objects (like numbers and sets of numbers) have an objective existence with properties that people can only discover, not invent, is generally ascribed to Plato and therefore is called Platonism. Gödel's adherence to this doctrine marked a clear shift in his views.

In a lecture given in Cambridge, Massachusetts in 1933, he had claimed that Platonism could not "satisfy any critical mind."[40] Researchers in set theory through the final decades of the twentieth century followed Gödel's injunction to seek new axioms, but despite much interesting work, the continuum hypothesis remains unsettled.

The most truly astonishing passage in Gödel's contribution to the Russell volume concerned Leibniz's pet project for a *universal characteristic.* Writing over two centuries after Leibniz's death, Gödel held out the hope that such a language could be developed and that it would revolutionize mathematical practice.

> But there is no need to give up hope. Leibniz did not in his writings about the *Characteristica universalis* speak of a utopian project; if we are to believe his words he had developed this calculus of reasoning to a large extent, but was waiting with its publication till the seed could fall on fertile ground. He went so far as to estimate the time which would be necessary for his calculus to be developed by a few select scientists to such an extent "that humanity would have a new kind of an instrument increasing the powers of reason far more than any optical instrument has ever aided the power of vision." The time he names is five years, and he claims that his method is not any more difficult to learn than the mathematics or philosophy of his time.[41]

Now, we have seen that what Leibniz had produced by way of a "calculus of reasoning," despite being amazing for its time, was a puny paltry thing compared with what Boole and Frege later accomplished. Whatever could Gödel have been thinking? Alas, it seems that he believed in a conspiracy to suppress Leibniz's ideas. Gödel had a number of very strange beliefs about many subjects, amounting to at least a touch of clinical paranoia. Yet his prestige among logicians is so great that there is hesitancy to simply dismiss any of his ideas. More about Gödel's mental problems later.

When Gödel was asked to write about Einstein for the *Library of Living Philosophers*, he chose as his topic the relationship between Einstein's relativity theory and Kant's philosophy. He found that the equations of the general theory of relativity (Einstein's theory of gravitation) possess a solution quite different from any physicists had imagined. Remarkably, Gödel's solution to these equations represents a universe in which a journey long enough and fast enough could end up in the past.

Naturally such a world is vulnerable to the paradoxes of time travel familiar to readers of science fiction: for example, could one travel to the past and kill one's own grandparent as a child? Gödel's surprisingly unphilosophical solution to this dilemma was to point out that such a voyage would be quite impractical if only because of the quantity of fuel required.

Gödel routinely revised his articles over and over again with meticulous care, withholding them from publication until he was completely satisfied. Even after publication, he would take the opportunity of a reprinting of one of his pieces to introduce further revisions. All in all this tended to be extremely frustrating to his editors watching deadlines recede. In the case of Gödel's promised essay on Rudolf Carnap for the *Library of Living Philosophers*, the volume finally appeared without his contribution.

However, six versions of his intended critique of Carnap's views on logic and mathematics were found among Gödel's papers, and the editors of his *Collected Works* decided to publish two of them. Another manuscript found among his papers was a handwritten draft (with various insertions, deletions, and footnotes) of the text of a lecture Gödel had given in Providence, Rhode Island, during Christmas week 1951.* In the lecture, entitled *Some Basic Theorems on the Foundations of Mathematics and Their Implications*, Gödel, in effect, placed Hilbert's dictum regarding the solvability of every mathematical question in the context of the nature of the human mind. Gödel raised the question of whether the human mind was in all essentials equivalent to a computer, a question still being vigorously debated in the context of prospects for artificial intelligence.

Without proposing to answer the question (although it ultimately became clear that he believed the correct answer is negative), Gödel maintained that either answer is "decidedly opposed to materialistic philosophy." If the full power of the human mind can be emulated by a finite mechanical device, then Gödel's own incompleteness theorem can be brought to bear to show that some proposition about the natural numbers, while true, can never be proved by human beings, an *absolutely undecidable* proposition.

This would contradict Hilbert's dictum. But according to Gödel, it would also require some measure of idealistic philosophy just to make sense of a statement that assumes the objective existence of natural numbers with properties beyond those that human beings can ascertain. On

*This was the prestigious once-a-year Gibbs lecture, given at the invitation of the American Mathematical Society. I was lucky to be in the audience at the lecture which had a profound influence on my own views about the foundations of mathematics.

the other hand, Gödel reasoned that if the human mind is not reducible to mechanism, whereas, as he believed was evident, the physical brain is so reducible, it would follow that the mind transcends physical reality, which again would be incompatible with materialism. It is not so much that this argument is totally persuasive, but rather that in bringing together considerations of theoretical logic, human physiology, the ultimate potential for computers, and fundamental philosophy, Gödel had once again shown his dazzling capacity to think in radically novel and unexpected directions.[42]

A Strange Man and a Sad End

As Kurt Gödel neared retirement age, he hoped that the logician Abraham Robinson, then at Yale, would take his place at the Institute for Advanced Study. Before any of this could happen, Robinson was diagnosed with inoperable pancreatic cancer, and died soon after. During his final months, Robinson received the following letter from Gödel:

> In view of what I said in our discussion last year [about Robinson coming to the Institute for an extended period of time] you can imagine how very sorry I am about your illness, not only from a personal point of view, but also as far as logic and the Institute for Advanced Study are concerned.
>
> As you know, I have unorthodox views about many things. Two of them would apply here:
>
> 1. I don't believe that any medical diagnosis is 100% certain.
> 2. The assertion that our ego consists of protein molecules seems to me one of the most ridiculous ever made.
>
> I hope you are sharing at least the second opinion with me. I am glad to hear that, in spite of your illness, you are able to spend some time in the mathematics department. I am sure this will provide some welcome diversion.[43]

This letter is quintessential Gödel. What he said about his distrust of medical diagnosis was certainly an understatement. When he suffered a total blockage of his urinary duct resulting from an enlarged prostate, he not only refused to accept the diagnosis, but also insisted that his problem could be treated with additional doses of the laxatives on which he had already become quite dependent. At one point he angrily ripped out the catheter that had been inserted. Refusing the surgery that usually relieves the blockage, he finally accepted the catheter and used it for the rest of his life.

His attempt to console Robinson by referring obliquely to his belief that the mind is more than "protein molecules" apparently with the suggestion that there would be an afterlife is another typical touch.

The boundary between Gödel's unorthodox views and outright clinical paranoia was not always clear-cut. Morgenstern records his surprise that Gödel took ghosts quite seriously. More important, Gödel was convinced that the refrigerators and radiators in his various apartments in Princeton were giving off noxious gases, as a result of which he and Adele moved a number of times. Finally he simply had the offending appliances removed, making his apartment "a pretty uncomfortable place in the winter time."

When Gödel sought to become a U.S. citizen, he prepared, in typical Gödel fashion, for the perfunctory examination on American institutions before a judge—he submitted the Constitution to the kind of meticulous analysis only he would have performed. Moreover, he became quite agitated when he concluded that the Constitution was actually inconsistent. While driving to Trenton, the state capital, for the process, Einstein and Morgenstern, his supporting witnesses, tried to distract Gödel from his "discovery," fearing it might cause trouble if broached. Einstein told one joke after another. But when the judge asked Gödel whether he thought a dictatorship like that in Germany was possible in the United States, the candidate began to explain his "discovery." Fortunately, the judge quickly understood with whom he was dealing and interrupted, so that all ended happily.

One may chuckle easily at such anecdotes revealing aspects of Gödel's strangeness. But it was not all so amusing. In a paranoid state over the safety of the food available to him, and with his devoted wife too ill to be much help, he literally starved himself to death. So, on January 14, 1978, ended the life of one of the great minds of the twentieth century.[44]

Appendix: Gödel's Undecidable Statement

The PM system is much too complicated to describe here. Instead the simpler PA system will be used to show some of the ingredients entering into the construction of undecidable propositions. PA can be set up using the 16 symbols

$$\supset\ \neg\ \vee\ \wedge\ \forall\ \exists\ \underline{1}\ \oplus\ \otimes\ x\ y\ z\ (\ )\ '\ \doteq .$$

Eccentric versions of the symbols 1, +, × and = have been used to emphasize that these are to be regarded as mere symbols, while at the same time suggesting their intended meaning. The letters x, y, and z are used as variables intended to range over the natural numbers. Because it is necessary to provide for more than three variables, the symbol $'$ is available to generate as many variables as one pleases by tacking it on to those letters. Thus y' and z''' are variables. Because there are more than 10 symbols, we'll use a coding scheme in which each symbol is replaced by a pair of decimal digits:

$$\begin{array}{cccccccccccccccc} \supset & \neg & \vee & \wedge & \forall & \exists & \underline{1} & \oplus & \otimes & x & y & z & (&) & ' & \doteq \\ \downarrow & \downarrow & \downarrow & \downarrow & \downarrow & \downarrow & \downarrow & \downarrow & \downarrow & \downarrow & \downarrow & \downarrow & \downarrow & \downarrow & \downarrow & \downarrow \\ 10 & 11 & 12 & 13 & 14 & 15 & 21 & 22 & 23 & 31 & 32 & 33 & 41 & 42 & 43 & 44 \end{array}$$

The natural numbers are represented by certain strings of these symbols called *numerals* as follows:

Numeral	**Number represented**	**Code**
$\underline{1}$	1	21
$(\underline{1} \oplus \underline{1})$	2	4121222142
$((\underline{1} \oplus \underline{1}) \oplus \underline{1})$	3	414121222142222142
$(((\underline{1} \oplus \underline{1}) \oplus \underline{1}) \oplus \underline{1})$	4	41414121222142222142222142
......	...	...

Most strings of the 16 symbols are just gibberish, for example:

$$\exists \oplus \otimes x \forall \neg \quad \text{or} \quad \doteq \supset \underline{1}'()$$

whose codes are 152223311411 and 441021434142, respectively. But certain of these strings, called *sentences*, can be used to express propositions, true or false, about the natural numbers. Thus, the string

$$((\underline{1} \oplus \underline{1}) \otimes (\underline{1} \oplus \underline{1}) \doteq (((\underline{1} \oplus \underline{1}) \oplus \underline{1}) \oplus \underline{1}))$$

whose code is

414121222142234121222142444141412122214222214222214242

expresses the true proposition that 2 times 2 is 4, while

$$((\underline{1} \oplus \underline{1}) \otimes (\underline{1} \oplus \underline{1}) \doteq ((\underline{1} \oplus \underline{1}) \oplus \underline{1}))$$

expresses the false proposition that 2 times 2 is 3. The sentence

$$(\forall x)(\neg(x \doteq \underline{1}) \supset (\exists y)(x \doteq (y \oplus \underline{1})))$$

whose code is

4114314241114131442142104115324241314441322221424242

expresses the proposition that every natural number except 1 has an immediate predecessor.

To complete our description of PA it would be necessary to specify certain sentences as *axioms* as well as the rules of inference to be used in proceeding from the axioms to the provable sentences. The list of steps along the way, beginning with axioms and ending with a sentence provable in PA, is called the *proof* of that sentence. Although to do this in full detail would take us too far afield, we consider as a simple example of what is involved, the sentence

$$(\forall x)\neg(\underline{1} \doteq (x \oplus \underline{1}))$$

which is intended to express the proposition that 1 is not the immediate successor of any natural number. This sentence might well be chosen as one of the axioms. Since sentences like this beginning with the symbol $\forall$ are intended to express assertions stating that some property holds for *all* natural numbers, one natural rule of inference that would apply to sentences of that kind would permit a substitution of some numeral for x (after removing the universal quantifier $(\forall x)$). This is just a matter of proceeding from a general statement to a specific instance of it. Here is a simple example:

$$\frac{(\forall x)\neg(\underline{1} \doteq (x \oplus \underline{1}))}{\neg(\underline{1} \doteq (\underline{1} \oplus \underline{1}))}$$

The conclusion, a provable sentence of PA, is obtained by substituting $\underline{1}$ for the variable x, and expresses the fact that 1 and 2 aren't equal.

In addition to strings that express propositions, there are others, called *unary*, that can be used to define sets of natural numbers. Such strings are to contain the symbol x but not the quantifiers $(\forall x)$ or $(\exists x)$ (although it may contain quantifiers with respect to other variables such as y or x''). In addition, unary strings are to possess the crucial property that if x is replaced everywhere by some numeral the resulting string is a sentence. An example of a unary string is:

$$(\exists y)(x \doteq ((\underline{1} \oplus \underline{1}) \otimes y))$$

whose code is

$$4115324241314441412122214223324242$$

If x is replaced by $(\underline{1} \oplus \underline{1})$ the true sentence

$$(\exists y)((\underline{1} \oplus \underline{1}) \doteq ((\underline{1} \oplus \underline{1}) \otimes y))$$

is obtained. If $\underline{1}$ is used instead, the false sentence

$$(\exists y)(\underline{1} \doteq ((\underline{1} \oplus \underline{1}) \otimes y))$$

is obtained. This unary string can be thought of as providing a definition of the set of even numbers. The more complicated unary string: string:

$$(\forall y)(\forall z)((x \doteq (y \otimes z)) \supset ((y \doteq 1) \vee (y \doteq x)))$$

whose code is

41143242411433424141314441322333424210414132441421241324431424242

defines the set consisting of 1 and all prime numbers. For a given unary string A and natural number n we'll use the notation $[A : n]$ to stand for *the sentence obtained by replacing x in A by the numeral that represents the number n*. For example,

$$[(\exists y)(x \doteq ((\underline{1} \oplus \underline{1}) \otimes y)) : 2]$$

stands for the sentence

$$(\exists y)((\underline{1} \oplus \underline{1}) \doteq ((\underline{1} \oplus \underline{1}) \otimes y)).$$

Now we can explain how Gödel's methods can be used to produce a sentence U of PA that expresses the proposition that it is not provable in PA. Using the code numbers assigned to unary strings, we can arrange all of them in order of the sizes of their codes. In this ordering the unary string with the smallest code is $(x \doteq \underline{1})$, and its code is 4131442142, over four billion. We write A_1 to stand for this unary string, and imagine all unary strings arranged in a sequence

$$A_1, A_2, A_3, \ldots$$

according to the size of their codes. Because these are unary strings, for any natural numbers n, m, the string $[A_n : m]$ will be a sentence. Some of these sentences will be provable in PA; others will not. For each n we can consider the set of those values of m for which $[A_n : m]$ *is not provable in* PA. Recalling our discussion of Cantor's diagonal method, we see that such a set can be thought of as a package with n as its label.

Applying the diagonal method, that is, identifying the label with one of the elements in the package it labels, we form the set K consisting of those numbers n such that $[A_n : n]$ is *not* provable in PA. The fact that provability in PA turns out to be definable in PA enables us to find (and, of course, this is the hard part) a unary string B that defines this very set K in PA. Now there must be some number q such that $B = A_q$ because all unary strings were included in the sequence of As. Thus, for every natural number n, the sentence $[A_q : n]$ expresses the proposition

$$[A_n : n] \textit{ is not provable in } \text{PA}.$$

In particular, with n being given the value q, we can see that $[A_q : q]$ expresses the proposition

$$[A_q : q] \textit{ is not provable in } \text{PA}.$$

So $[A_q : q]$ is a sentence of PA that expresses the proposition that it is not provable in PA.

CHAPTER 7

Turing Conceives the All-Purpose Computer

As early as 1834, Charles Babbage had conceived an automatic calculating machine. His proposed but never constructed *analytical engine* was intended to carry out numerical computations of the most varied kind.* To emphasize the power and scope of his engine, Babbage remarked facetiously that "it could do everything but compose country dances."[1] While for Babbage, it was self-evident that machines designed for computation could not be expected to compose dances, it does not strike us today as being at all out of the question. In fact, today's computers can perfectly well be programmed to compose country dances (although perhaps not of the finest quality). Someone today seeking a similar figure of speech

Computers can do everything but ...

to emphasize the power and scope of computers would not find it easy to complete the sentence. Almost any imaginable task involving symbols, numbers, or text is already within the competence of computers or some expert is insisting that it soon will be so. As a last resort one could try "Computers can do everything but read our thoughts," or "Computers can do everything but communicate with angels."

Clearly, our very concept of what constitutes computation has been altered drastically. The underlying conception on which this expanded view of computation is based was formulated by Alan Turing in 1935 in the process of solving a problem in mathematical logic posed by David Hilbert.

Babbage had intended to build his engine entirely out of mechanical components like gears, and given the complexity of the proposed device, it is not surprising that he failed. It was only with the development, beginning in the 1930s, of electromechanical calculators using electrical relays,

*Charles Babbage was born in London in December 1792. An accomplished mathematician, he was part of a group seeking to bring continental mathematical ideas to the British universities. He developed a particular interest in mechanical calculation and conceived a "difference engine," designed for the efficient construction of mathematical tables. Soon Babbage was inspired to propose his far more ambitious analytical engine. He died in 1871 an embittered man, frustrated over the failure to complete this project.

that machines achieved the scope Babbage had envisioned. But during the 1930s and 1940s none of those involved with this work spoke of machines going beyond straightforward mathematical calculation. As we will see, the person who first succeeded in bringing Babbage's vision to life was Howard Aiken. He wrote:

> If it should turn out that the basic logics of a machine designed for the numerical solution of differential equations coincide with the logics of a machine intended to make bills for a department store, I would regard this as the most amazing coincidence that I have ever encountered.[2]

Aiken made this remarkable assertion in 1956 when computers that could readily be programmed to do both of these things were already commercially available. If Aiken had grasped the significance of Alan Turing's paper, published two decades earlier, he would never have made such a preposterous statement.

A Child of the Empire

Alan Turing's father, Julius Turing, was a great success as a civil servant in India. In the spring of 1907, after more than a decade of service, he was ready for a leave to England. It was on the voyage home, via the Pacific, that he met Alan's mother, Ethel Sara Stoney. She had been born in Madras, and after growing up in Ireland and spending six months in Paris, had returned to India. A shipboard romance developed swiftly, and they crossed the United States together, stopping for a tour of Yellowstone Park. With her father's approval, they married in Dublin in the fall, before returning to India that winter.

Alan's older brother John was born in September 1908. Julius's duties required extensive travel in the south of India, often accompanied by Ethel Sara and the baby. It was in the fall of 1911, while on these rounds, that Alan was conceived. After Julius managed to obtain another leave, the family sailed together to England. Alan Mathison Turing was born in London on June 23, 1912.[3]

The remorseless work required to manage the empire made a life together a difficult proposition for the Turing family. The father's career was in India where the prevalence of tropical disease was particularly dangerous for young children, and where a befitting education for them was not to be had. The mother could be with her husband or with her children; only when the father was on leave could she be with both. Alan was only 15-months old when his mother arranged to leave him and his four-year-old brother to board with a retired colonel and his wife in England, while she returned to India.

Alan Turing
(© National Portrait Gallery, London)

Mrs. Turing managed to spend a few months with the children in 1915, and in the spring of 1916 both parents made the voyage home. But this time the trip was dangerous because of the German submarines, and so Mrs. Turing remained in England when her husband returned to India. Thus the grim war actually benefited Alan by keeping his mother in England. He was a precocious happy child who made friends readily, but was clumsy and untidy. It was not at all uncommon for six-year-old boys to be sent to boarding schools, but Alan's mother kept him at home, sending him to a local day school to learn the Latin deemed essential. There he struggled with scratchy pens, leaky fountain pens, and terrible penmanship.

When his mother left for India in 1919, the seven-year-old Alan returned to the colonel's establishment. After an interval of almost two years, she returned to find that her child had not been doing very well. Instead of the cheerful little boy she had left, she found an "unsociable" introverted child whose basic education had been badly neglected. After doing her best to get him ready, she enrolled him in the small boarding school where his brother John was already a student. The two were together only for a few months before John left for a "public school."* So it was, that after a summer vacation, Alan was left by his parents to cope alone with boarding school life. He showed what he felt about the prospect by running miserably after his parents' departing automobile.

By the time the 14-year-old Alan Turing began his residence at the Sherborne public school, his passion for science and mathematics had been established. He found himself in an environment in which competitive sports were valued and mathematics most emphatically was not.* One of his teachers thought of science in general as "low and cunning," and spoke of mathematics as imparting a bad smell to a room.[4]

Alan's mathematical genius was recognized but belittled. His parents were warned of the danger of his becoming a mere science specialist. Above all, there was the dirty, blotted work in his almost illegible handwriting. Meanwhile, having little to do with the other boys and paying little attention to his classes (but doing well enough on exams), Alan carried on his own little mathematical investigations and studied Einstein's theory of relativity.

Life changed for Alan when he found a friend, and more than a friend. Christopher Morcom shared Alan's passion for science and mathematics. Unlike Alan, Christopher was a diligent student who took all of his school

*As most readers probably realize, the British "public" schools are in fact elite private institutions. Attendance at one was a crucial milestone on a boy's journey towards a successful upper-middle-class career.

*Wellington, the victor of the Battle of Waterloo, is supposed to have said that battle was won "on the playing fields of Eton." Eton was regarded as the most elite of the public schools.

work seriously, and whose written work was impeccably neat. Alan's admiration for Chris knew no bounds, and he determined to be more like him.

It is unclear at what point in his life Alan Turing became fully aware of his homosexuality, but it is natural to suppose that, at least for Alan, the friendship with Chris had erotic overtones. Turing's biographer calls Alan's feelings "first love," and indeed they had that intensity. It is impossible to know how the relationship would have developed, how Alan's feelings might have been modulated, had not tragedy intervened. Unbeknownst to Alan, his friend had been suffering from tuberculosis, and he died in February 1930, remaining forever enshrined as a symbol of perfection in Alan's mind.[5]

By his final year at Sherborne, Alan had become so successful in his studies that he was able to win a scholarship to Kings College at Cambridge University. In addition to room and board, he was provided with a stipend of £80 a year, just about half of what a skilled worker could have hoped to earn at the time.[6] Where mathematics had been in bad odor at Sherbourne, at Cambridge Turing found himself in an atmosphere in which his mathematical genius could flourish.

Cambridge's great mathematician was G. H. Hardy (1877–1937) whose *Course of Pure Mathematics*, published in 1908 has been a classic textbook from which successive generations of student mathematicians came to grips with the fundamental properties of limit processes. (It is still in print as this is being written.)

Hardy has been portrayed in connection with a self-educated postal clerk from Madras named Ramanujan, whose mathematical genius Hardy brought to light, both in a public television program and in a popular film.

Among the lecture courses available to Turing were those of Hardy and also those of the mathematical physicist and astronomer Sir Arthur Eddington, who had led the 1919 expedition to West Africa, where a total eclipse of the sun had made it possible to observe the behavior of starlight passing near the sun, and thereby obtain the first confirmation of Einstein's prediction, in his general theory of relativity, concerning the bending of such light as the result of the sun's gravitational pull.

Eddington's lectures raised the question of why so many statistical observations seemed to align themselves along the famous bell-shaped curve known as the "normal" distribution. Eddington's lectures would also have covered the still new quantum theory then revolutionizing physics. But the work in this area that attracted Turing's serious attention was a recently published book on the mathematical foundations of quantum mechanics by John von Neumann (whom we have met in the previous two chapters and will encounter again), a book Alan had won as a prize at Sherborne.

The ubiquitous occurrence of the bell-shaped normal distribution stressed in Eddington's lectures fascinated Turing, and he sought an underlying mathematical explanation. He found it by working out a proof that a diverse range of statistical distributions do tend "in the limit" to the normal distribution. This was an application par excellence of the limit processes of the calculus. Alan Turing didn't know that he had discovered nothing new, that his result was well known as the "central limit theorem." Nevertheless, his achievement was found sufficiently impressive for him to be offered a position as a fellow, although ordinarily this would have required something new.

Turing was now a Cambridge "don" with an annual stipend of £300. It was a three-year appointment with an almost automatic renewal for another three. There were no specific duties attached to the fellowship, and Turing could now take dinner at "high table," literally looking down on the undergraduates. If he wished, he could earn additional income by serving as a tutor to undergraduates. The appointment placed Turing on a path ordinarily expected to lead to an academic career.*

At Sherborne, the alleged bad smell of mathematics and the admonitions to avoid being a mere "scientific specialist" were forgotten amidst the celebration of the success of one of their old boys. The students were granted a special half-day holiday, and the verse

> Turing
> Must have been alluring
> To get made a don
> So early on.[7]

was shamelessly passed around.

It didn't take long after his appointment for Turing to produce his first bit of genuinely new mathematics leading to a published paper. As it happened, what he had obtained was an improvement of a theorem proved by von Neumann in a highly specialized field known as the theory of "almost-periodic functions." Turing was now well on the way to a career as a successful research mathematician whose accomplishments would be of interest only to other specialists. Then, he attended a course of lectures on the foundations of mathematics given at Cambridge in the spring of 1935, and Turing learned about the Entscheidungsproblem.

Hilbert's *Entscheidungsproblem*

Leibniz had dreamt of human reason reduced to calculation and of powerful mechanical engines to carry out calculations. Frege had provided for the

*The doctorate, a standard requirement for a university appointment in France, Germany, and the United States, was rarely sought by English academics before the Second World War.

first time a system of rules that could plausibly account for all of human deductive reasoning. Gödel, in his doctoral dissertation of 1930, had proved that Frege's rules were complete, answering a question posed by Hilbert two years earlier. Hilbert had also sought explicit calculational procedures by means of which it would always be possible to determine, given some premises and a proposed conclusion, written in the notation of what has come to be called "first-order logic" whether Frege's rules would enable that conclusion to be derived from those premises.[8]

The task of finding such procedures came to be known as Hilbert's *Entscheidungsproblem* (literally: decision problem), Of course, systems of calculational procedures for solving specific problems were not new. Indeed the traditional mathematical curriculum has been largely made up of such calculational procedures, otherwise known as *algorithms*. We begin by learning algorithms for addition, subtraction, multiplication, and division of numbers, we move on to algorithms for manipulating algebraic expressions and solving equations, and, if we continue to calculus, we learn how to use the algorithms originally developed by Leibniz for that subject.

However, Hilbert was asking for an algorithm of unprecedented scope. In principle, an algorithm for his *Entscheidungsproblem* would have reduced all human deductive reasoning to brute calculation. To a considerable extent, it would have been a fulfillment of Leibniz's dream.

Mathematicians often like to approach a difficult problem from two directions. On the one hand, they try to do what they can with special cases of the general problem. Moving in the other direction, they try to reduce the general problem to certain special cases. If all goes well, the two approaches meet in the middle, providing a solution to the general problem.

Work on the *Entscheidungsproblem* proceeded on exactly these lines, and indeed the gap between the special cases for which algorithms had been found and the cases to which the general problem had been reduced had been narrowed to such an extent that it was possible to hope that a small further advance would eliminate the gap entirely, and thus provide the algorithm Hilbert sought.[9] One skeptic was Cambridge's G. H. Hardy who somewhat indignantly commented: "There is of course no such theorem, and this is very fortunate, since if there were we should have a mechanical set of rules for the solution of all mathematical problems, and our activities as mathematicians would come to an end."[10] Hardy was certainly not the first craftsman to be convinced that his skill could never be replaced by a mere mechanism, but this craftsman turned out to be right!

Another Cambridge don, M. H. A. (Max) Newman, 15 years older than Turing and a Fellow of St. John's College, was to play an important and continuing role in the younger man's career. Newman had made pioneering contributions to topology, at the time a relatively new branch of mathematics. Roughly speaking, topology deals with properties of geometric figures that remain undisturbed by any amount of stretching, so long as there is no tearing.

Newman's lecture course on topology at Cambridge introduced many young mathematicians to this burgeoning field, and he wrote an excellent textbook on the subject. When Newman attended the 1928 International Congress of Mathematicians in Bologna, he heard Hilbert set forth goals that, only two years later, the young Kurt Gödel would show were unattainable. Apparently intrigued by these developments, Newman gave a lecture course in the spring term of 1935 on the foundations of mathematics featuring Gödel's incompleteness theorem as its climax. Attending this course, Turing learned about Hilbert's *Entscheidungsproblem*. Quite apart from the incredulity of mathematicians such as Hardy, after Gödel's work it was hard to believe that there could be an algorithm such as Hilbert had wanted. Alan Turing began to think about how it could be possible to *prove* that no such algorithm exists.

Turing's Analysis of Computation Process

Turing knew that an *algorithm* is typically specified by a list of rules that a *person* can follow in a precise mechanical manner, like a recipe in a cookbook. But he shifted his focus from the rules to what the person actually *did* when carrying them out. He was able to show, by a process of successively stripping away inessential details, that such a person could be limited to a few extremely simple basic actions without changing the final outcome of the computation.

Turing's next step was to see that the person could be replaced by a machine capable of performing these same basic actions. Then, by proving that no machine performing only those basic actions could determine whether a proposed conclusion follows from given premises using Frege's rules, he was able to conclude that no algorithm for the *Entscheidungsproblem* exists. As a byproduct, he found a mathematical model of an all-purpose computing machine.

To try to follow what Turing's thought processes might have been, let us imagine ourselves watching a computation in progress. What was the person doing the computing actually doing? She (for it seems that most often women did this work in the 1930s) was making marks on a sheet of paper.* She could be observed shifting her attention back and forth from what she had written earlier to what she is writing now.

Turing wanted to strip this description of irrelevant detail. Was she sipping a cup of coffee while working? Surely not relevant. Was she writing

*In fact at this time, "computer" meant a person (typically female) whose job was performing computations.

with pencil or with pen? Again, that surely doesn't matter. What about the size of the sheets of paper? Well, if the paper size is small, she may well need to look back at previous sheets more often. But Turing easily convinced himself that this was a matter of convenience not of necessity. Nothing essential would really change if she were restricted to paper so short that she couldn't write symbols under one another, in effect if she used something like a roll of paper tape ruled into horizontal squares. To keep things simple, let us imagine that she is working out a multiplication example:

$$\begin{array}{r} 4231 \\ \times 77 \\ \hline 29617 \\ 296170 \\ \hline 325787 \end{array}$$

Without losing anything essential, we can imagine her doing her work along a paper tape like this:

4	2	3	1	×	7	7	=	2	9	6	1	7	+	2	9	6	1	7	0	=	3	2	5	7	8	7

Turing convinced himself that, while it might be a bit of a nuisance to deal with a complicated calculation along such a one-dimensional tape, there was no fundamental problem in doing so. Let us continue to observe the computation in progress, now restricted to a roll of paper tape. We watch as our subject glances back and forth along the tape, writing symbols, sometimes backing up and erasing symbols so new symbols can be written in their place. Her decision about what to write next will depend on which symbols she is paying attention to, but also on her current *state of mind.* Even in the case of our simple multiplication example, as she notes pairs of digits, her state of mind will determine whether she multiplies or adds them. As she begins, her tape looks like this:

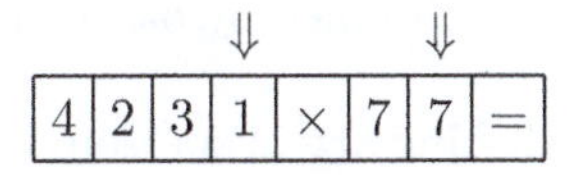

An arrow (⇓) appears above the digits 1 and 7 to indicate that those symbols are initially receiving her attention. Multiplying them gives 7 which she writes on the tape:

		⇓				⇓		
4	2	3	1	×	7	7	=	7

She has now shifted her attention to the digits 3 and 7 which are to be multiplied in turn. After the phase of the computation in which she multiplies pairs of digits is completed, she will need to add her two partial products:

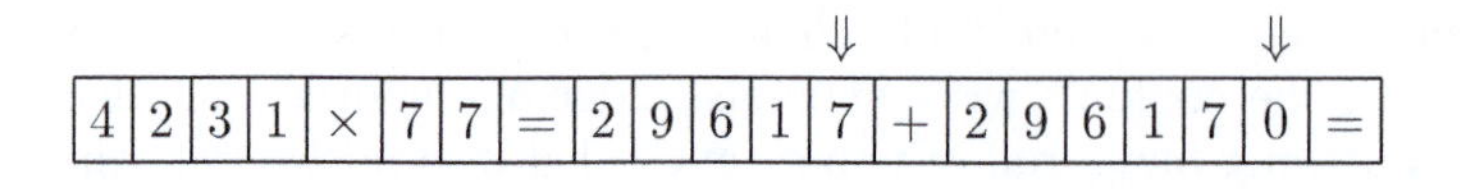

She begins this phase by adding 7 and 0, obtaining:

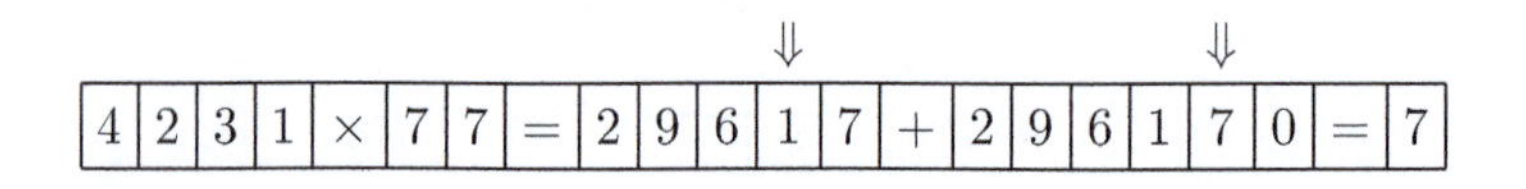

Now she must add 1 and 7 to get 8. Notice that the digits that have her attention at this point, 1 and 7, are the same digits that she multiplied when she began the calculation. But although the digits are the same, her *state of mind* is different and leads her to add them instead.

The simple example illuminates crucial features of any computation. A person carrying out a computation—in arithmetic, algebra, calculus, or any other branch of mathematics—operates under the following constraints:

- At each stage of a computation only a small number of symbols receive attention.
- The action taken at each such stage depends only on the particular symbols receiving attention and on the current state of mind of the person carrying out the computation.

How many symbols can a person deal with simultaneously? And how many are really needed to carry out a computation correctly? As for the first question, the number surely depends on the particular person, but in any case the answer is not very many. Regarding the second question, the answer is 1. This is because the effect of paying attention to several symbols simultaneously can always be obtained by paying attention to each of them singly.[11]

Moreover the effect of shifting attention from a particular square on the tape to another a certain distance away can be obtained by a succession of moves, each of which involves a shift one square to the right or one to the left. This analysis leads to the conclusion that any computation whatever can be envisioned as proceeding in the following manner:

- The computation is carried out by writing symbols in squares on a ruled paper tape.

- At each step the person performing the computation pays attention to the symbol written in only one of these squares.
- Her next action will depend on this symbol and her state of mind.
- This next action will consist of writing a symbol on the square to which she has been paying attention and then possibly shifting her attention to the square immediately to the left or immediately to the right.

Now it is easy to see that the person doing the work can be replaced by a machine: the tape—which can be visualized as magnetic tape with the written symbols represented by coded information along the tape—moves back and forth in the machine. The states of mind of the person carrying out the computation are represented by different configurations of the internals of the machine. The machine must be designed so that at each moment it is sensitive to exactly one symbol on the tape, the *scanned symbol.*

Depending on its internal configuration and on the scanned symbol, the machine will write a symbol on the tape (replacing the one scanned) and then either continue to scan the same square, or else shift to the position immediately to the left or to the right on the tape. For the purpose of the computation, it doesn't matter how the machine is constructed or even what it is made of; all that is significant is that it have the capability of assuming a number of different *configurations* (also called *states*) and that it behave appropriately in each such configuration or state.

The point is not to actually build one of these so-called *Turing machines*—after all, they are merely mathematical abstractions.* What is important is that on the basis of Turing's analysis of the notion of *computation*, it is possible to conclude that anything computable by any algorithmic process can be computed by a Turing machine. So if we can prove that some particular task cannot be accomplished by a Turing machine, we can conclude that no algorithmic process can accomplish that task. That is how Turing proved that there is no algorithm for the *Entscheidungsproblem.* In addition, Turing showed how to produce a Turing machine that alone could do anything that could be done by any Turing machine whatever—a mathematical model of an all-purpose computer.

Turing Machines in Action

Turing's analysis of the process of computation has led to the conclusion that any computation can be carried out by one of the severely circumscribed devices that have come to be called Turing machines. It will be worthwhile to examine a few simple examples.

*Turing called his abstract inventions *a*-machines—"*a*" for "automatic."

What is needed to exhibit a particular Turing machine? To begin with, a list of all of its possible states is required. Then, for each of these states and each symbol that might be encountered on the tape, it is necessary to specify the machine's action in that state when confronted by that symbol. This action, let it be recalled, is to consist simply of a possible change of symbol on the square being scanned, a movement one square to the left or to the right, and a possible change of state. Using uppercase letters to stand for the different machine states, we can symbolize the statement:

> *When the machine is in state* R *scanning the symbol* a *on the tape, it will replace* a *by* b, *move one square to the right, and then shift into state* S

by the formula R a : b → S. The analogous statement calling for motion one square to the left will similarly be symbolized as R a : b ← S. Finally, a statement calling for a change of the symbol on the tape without any motion along the tape will be symbolized as R a : b ⋆ S. It is usual to call these formulas *quintuples* because it takes five symbols to specify one of them (not counting the colon). Any Turing machine may then be characterized by a list of such quintuples.

Let us see how to produce a Turing machine that tests a given natural number to see whether it is even or odd. The given number will be written in the familiar (decimal) notation as a string of the digits 1, 2, 3, 4, 5, 6, 7, 8, 9, 0. Of course it is very easy to tell at a glance whether a number (written this way) is even or odd. Just look at the *rightmost* digit: if it is 1, 3, 5, 7, or 9, the number is odd; otherwise it is even. But the setup we'll use will have the machine beginning by scanning the *leftmost* digit. Since a Turing machine can only deal with one digit at a time and can only move one square at a time, how to manage this computation isn't totally obvious. The "input" number is written on the tape like this:

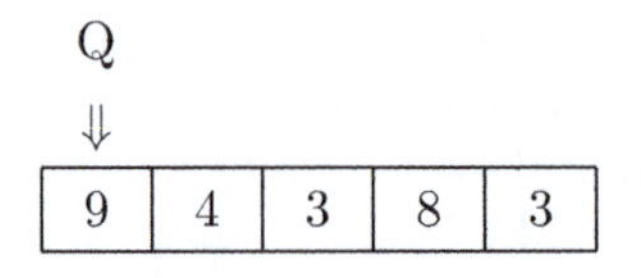

Here 94383 is written on the tape, and the machine is shown in its initial state Q scanning the leftmost square. Although the tape is shown with only five squares (just enough to contain the input) it is crucial that there be no limit to the amount of tape available for a computation. For this reason a blank square will always appear if the machine tries to move off the right end of the tape. We treat the blank as a special character, written □.

Our Turing machine will always start in state Q scanning the leftmost square. Whatever number is input to the machine, it will eventually terminate with a tape that is blank except for one square. That one square

Q 0 : □ → E	Q 2 : □ → E	Q 4 : □ → E	Q 6 : □ → E	Q 8 : □ → E
Q 1 : □ → O	Q 3 : □ → O	Q 5 : □ → O	Q 7 : □ → O	Q 9 : □ → O
E 0 : □ → E	E 2 : □ → E	E 4 : □ → E	E 6 : □ → E	E 8 : □ → E
E 1 : □ → O	E 3 : □ → O	E 5 : □ → O	E 7 : □ → O	E 9 : □ → O
O 0 : □ → E	O 2 : □ → E	O 4 : □ → E	O 6 : □ → E	O 8 : □ → E
O 1 : □ → O	O 3 : □ → O	O 5 : □ → O	O 7 : □ → O	O 9 : □ → O
E □ : 0 ⋆ F	O □ : 1 ⋆ F			

The set of quintuples constituting the Turing machine

will contain a 0 if the original input was even and a 1 if it was odd. The machine will have four states symbolized by Q, E, O, and F. As stated Q is the initial state.

Whatever state the machine is in, if it scans an even digit it will erase that digit (i.e., it will print a blank over it), move one square to the right, and then enter state E. Similarly, scanning an odd digit, it will erase it, move to the right, and enter state O. Eventually it will have scanned and erased the entire input and arrive at an empty square. At this point it will print a 0 if it has arrived in state E and a 1 if it has arrived in state O. Then, it will move one square to the left and halt. The set of quintuples constituting this machine is shown in the illustration on the previous page.

The complete computation beginning with the sample input 94383 shows the operation of the machine in detail:

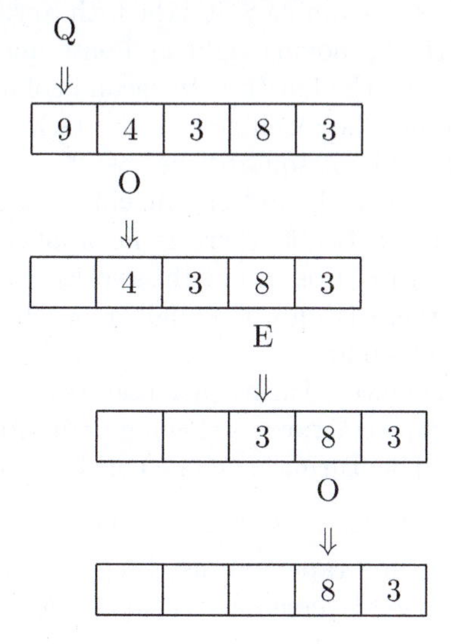

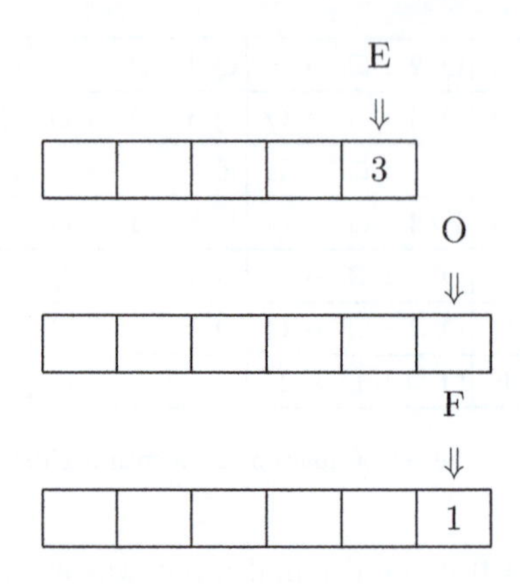

The machine begins in state Q scanning the 9. The applicable quintuple is in the second row, last column of the listing. This quintuple causes the machine to erase the 9, move one place to the right, and to enter the state O. In state O scanning a 4, the quintuple in the fifth row, third column applies.

Accordingly, the machine erases the 4, moves to the right and enters state E. Next, in state E, scanning a 3, the quintuple in the fourth row, second column causes the 3 to be erased and the machine continues right, entering state O. In O, scanning 8, it's the fifth row, last column that comes into play, erasing the 8, moving right and entering state E. Once again in state E scanning 3, it's the fourth row, second column that applies causing the 3 to be erased and the machine to move right entering state O.

In O scanning a blank square, the last row, second column applies. The blank is replaced by 1, and the machine stays put and enters state F. In state F, facing a blank, there is no applicable quintuple, and the machine (doubtless tired from all of this work) halts. At the conclusion of the computation, there is only the digit 1 on the tape, which is correct because the input was odd.

Unlike physical devices, Turing machines benefit from their existence as mere mathematical abstractions by having no limitations on the amount of tape they can use. The Turing machine consisting of the single quintuple

$$Q\,\square : \square \rightarrow Q$$

when started on a blank tape will just keep moving to the right "forever" as the amount of tape traversed keeps expanding:

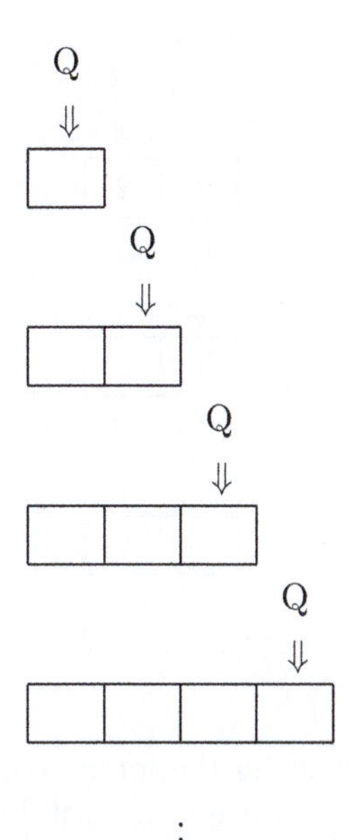

⋮

A Turing machine computation can continue "forever" without ever halting even when it only traverses a fixed amount of tape. For example, consider the Turing machine consisting of the two quintuples:

$$\text{Q}\ 1 : 1 \rightarrow \text{Q} \qquad \text{and} \quad \text{Q}\ 2 : 2 \leftarrow \text{Q}.$$

With input 12, this machine will bounce back and forth "forever" like this:

Q
⇓

1	2

      Q
      ⇓

1	2

Q
⇓

1	2

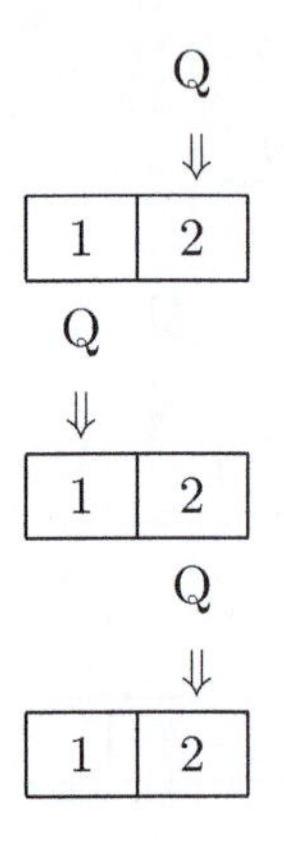

This behavior is dependent on the input. For example, if the input is 13, the computation of that same machine will be as follows:

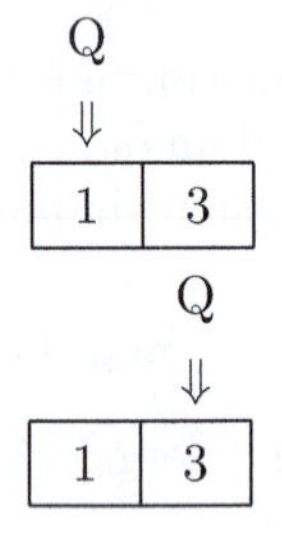

In state Q scanning 3, no quintuple is applicable and the machine halts.

In summary, some Turing machines with some inputs eventually halt; others do not. Applying Cantor's diagonal method to this situation led Turing to problems that could not be solved by Turing machines, and from there to the unsolvability of the *Entscheidungsproblem.*

Turing Applies Cantor's Diagonal Method

The climax of Max Newman's course that had brought the *Entscheidungsproblem* to Alan Turing's attention was Gödel's incompleteness theorem. So it was natural for Turing, contemplating the representation of his machines as lists of quintuples, to think of using natural numbers as codes for the machines, and of using Cantor's diagonal method. We'll follow Turing's line of thought, and set up a code similar to, but not identical with, the one he used.

For the purpose of setting up our coding scheme, we think of the quintuples constituting a Turing machine written one after the other separated by semicolons. Thus the Turing machine consisting of the pair of quintuples:

$$\text{Q } 1 : 1 \rightarrow \text{Q} \qquad \text{Q } 2 : 2 \leftarrow \text{Q}$$

would be written: Q1:1→Q;Q2:2←Q. Then we replace each symbol by a string of decimal digits according to the following scheme:

- Strings beginning and ending with 8 with only the digits 0, 1, 2, 3, 4, 5 between will be used for symbols on the tape. The table below gives the representations we'll use for the decimal digits and □ (as tape symbols) and also for the symbols → ← ⋆ : ;

Symbol	**Representation**	**Symbol**	**Representation**
0	8008	□	8558
1	8018	→	616
2	8028	←	626
3	8038	⋆	636
4	8048	:	646
5	8058	;	77
6	8518		
7	8528		
8	8538		
9	8548		

- Strings beginning and ending with 9 with only the digits 0, 1, 2, 3, 4, 5 between will be used for *states*. In particular, the start state Q will be represented by the string 99.

Thus the two-quintuple Turing machine would be coded by 998018 646 8018 616 99 77 998028 646 8028 626 99. For the Turing machine we built to distinguish even from odd numbers, we can code the states E, O, F by 919, 929, and 939, respectively. The code number for the entire machine would then be:

998008646855861 6919 77 9980286468558616919 77 9980486468558616919 77
9985186468558616919 77 9985386468558616919 77 9980186468558616929 77
9980386468558616929 77 9980586468558616929 77 9985286468558616929 77

9985486468558616929 77 91980086468558616919 77 91980286468558616919 77
91980486468558616919 77 91985186468558616919 77 91985386468558616919 77
91980186468558616929 77 91980386468558616929 77 91980586468558616929 77
91985286468558616929 77 91985486468558616929 77 92980086468558616919 77
92980286468558616919 77 92980486468558616919 77 92985186468558616919 77
92985386468558616919 77 92980186468558616929 77 92980386468558616929 77
92980586468558616929 77 92985286468558616929 77 92985486468558616929 77
91985586468008636939 77 92985586468018636939

Although this is just one big number, it has been displayed with spaces to show the codes for the individual quintuples. Notice that it is a straightforward matter to recover the quintuples from the code: First find the 77s that separate the codes of the individual quintuples, and then decode each quintuple. For example, the code 92985386468558616919 separates into 929 8538 646 8558 616 919 which decodes into O 8: □ → E. Of course the coding could have been set up in many different ways, but this scheme has this important and useful property of transparent decodability.*

As with the above examples, any Turing machine can be thought of as initially scanning the leftmost digit of a number written on its tape. For some of these numbers, the machine will eventually halt, while for others it may continue forever. Let us call the set of those natural numbers in the first of these categories the *halting set* of that particular Turing machine. Now, if we think of the halting set of a Turing machine as constituting a *package and of the code number of that machine as labeling that package,* then we have a typical setup for applying the diagonal method: labeled packages in which the labels are exactly the kinds of things in the packages—in this case, natural numbers.† The diagonal method will permit us to manufacture a set of natural numbers we will call D that is different from any halting set of a Turing machine. Here's how:

> *D will consist entirely of code numbers of Turing machines. For each Turing machine, its code number will belong to D if and only if it does not belong to the halting set of that machine.*

Thus, if the code number of some particular Turing machine belongs to its halting set, then that code number doesn't belong to D. On the other hand, if that code number doesn't belong to the machine's halting set, then

*Note that this coding scheme allows for symbols on the tape other than the decimal digits and □, symbols coded by such strings as 81118. This allows for symbols that can mark particular squares on the tape so they can be found on a return visit. It is possible to prove that the use of such additional symbols does not increase the computational power of Turing machines. It can also be proved that the use of the decimal system is irrelevant to what Turing machines can do. (Davis et al., 1994, pp. 113–168).

†For a quick refresher on this see pp. 57–59 of this book.

it does belong to D. In either case D cannot be the same set of numbers as the halting set of the machine in question. Since this is the case for every Turing machine, we can conclude:

The set D is not the halting set of any Turing machine.

But wait! We present a stubborn person who remains unconvinced. We listen in on a conversation between the Stubborn Person (SP) and the Omniscient Author (OA):

SP: I didn't quite follow that reasoning, but in any case I know that I can construct a Turing machine whose halting set is D. In fact (showing a piece of paper) here it is.

OA: I see. Would you kindly calculate the code number of your machine.

SP: Gladly! Let me see. The number is

998038646855861692977......7792985286468558616929

(showing some enormous number).

OA: OK. And is this number in the halting set of your machine?

SP: Wait! I must work this out. No. No. It's not in my machine's halting set.

OA: Now listen, Stubborn Person! If this number is not in your machine's halting set, then from the way D was defined, the number must be in D. Since this number is in D and is not in your machine's halting set, the two sets must be different.

SP: Let me check my work. Oh I see. I made a small mistake. Very silly of me. In fact this number is in my machine's halting set. I apologize for my foolish mistake.

OA: Not so fast! From the way D was defined, if the code number of your machine is in its halting set, then it most certainly is not in D. So the two sets must be different.

SP: What you are saying sounds plausible enough. But if I were to agree that you'd proved your point, then I'd no longer be a Stubborn Person.

Unsolvable Problems

A set of natural numbers D has been defined that is different from the halting set of any Turing machine. But what possible connection can this possibly have to the *Entscheidungsproblem*? The connection has to do with the reason that Hilbert called this *the fundamental problem of mathematical logic*. Hilbert understood that a solution to the *Entscheidungsproblem* would provide an algorithm for settling all mathematical questions. This same understanding underlay Hardy's certainty that there would never be a solution to the *Entscheidungsproblem*. If we take this seriously, it implies that if there is any example of a mathematical problem which can be shown to be algorithmically unsolvable, then the *Entscheidungsproblem* must be unsolvable. The set D will provide us with such an example.

We consider the following problem:

> Find an algorithm to determine for a given natural number whether or not it belongs to the set D.

This is our example of an unsolvable problem. Our first step in showing that there is no such algorithm is to observe that by Turing's analysis of the computation process, if there were such an algorithm, there would be a Turing machine that could accomplish the same thing. Just as with the Turing machine constructed to distinguish even from odd numbers, we can visualize such a machine as beginning to scan the leftmost digit of the given number in an initial state Q, like this.

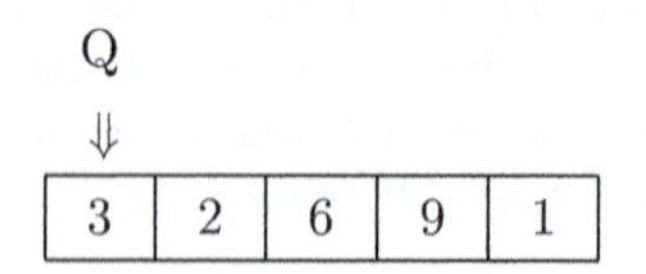

Likewise, we would want the machine to halt eventually, with a tape that is blank except for a single digit: 1 if the input number belongs to the set D, and 0 if it doesn't. Finally we would want it to halt in a state F with the property that no quintuples of the machine begin with the letter F.* For example,

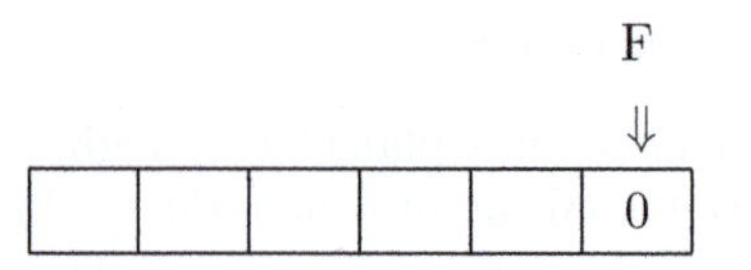

*It should be emphasized that if there really were an algorithm for distinguishing members of D from non-members, there would be no problem with these input-output embellishments. After all, there would be no difficulty with handing the input number to a person to execute the supposed algorithm in that form, nor would there be a problem in having the person put the output on the tape in the desired form.

Now let us imagine adding the following two quintuples to our imagined Turing machine:

$$\mathrm{F}\ 0 : \square \to \mathrm{F} \quad \text{and} \quad \mathrm{F}\ \square : \square \to \mathrm{F}.$$

With an input that belongs to D, the new machine will behave as before, eventually coming to a halt with 1 on the tape. However, with an input that doesn't belong to D, this machine will move to the right forever. Hence, the halting set of this supposed new machine is exactly the set D. However, this is impossible because D was constructed using the diagonal method so as to be different from the halting set of any Turing machine whatsoever. So our supposition that there is an algorithm for distinguishing members from non-members of D must have been wrong. There is no such algorithm! The problem of algorithmically distinguishing members from non-members of D is unsolvable!

As we have seen, Hilbert and Hardy both believed that an algorithmic solution to the *Entcheidungsproblem* would imply that any mathematical problem can be decided by an algorithm. So once we have a mathematical problem that is algorithmically unsolvable, the unsolvability of the *Entscheidungsproblem* should follow. To see how to make the connection with the set D, we associate with each natural number n the following proposed premise and conclusion:

PREMISE. The natural number n is the code number of some Turing machine and the same number n is placed on its tape with the leftmost digit scanned.

CONCLUSION. A Turing machine started in that manner will eventually halt.

Using the language of first-order logic both of these sentences can be translated into logical notation. It is then possible to prove that the conclusion can be derived using Frege's rules from the premise if and only if the Turing machine in question really will eventually halt when started with its own code number on its tape. And, this in turn is true if and only if n does not belong to D.

So, if we possessed an algorithm for the *Entscheidungsproblem*, we could use it to decide membership in D. Namely given a natural number n we could use our supposed algorithm for the *Entscheidungsproblem* to check whether or not the conclusion follows from the premise. If it does, we would know that n doesn't belong to D, and if not, we would know that n does belong to D. It follows that the *Entscheidungsproblem* is algorithmically unsolvable.[12]

Turing's Universal Machine

There was something troubling about what Turing had done. He had proved that no Turing machine could be used to solve the *Entscheidungsproblem.* However, to conclude that there is no algorithm of any kind for the *Entscheidungsproblem,* Turing had recourse to his discussion of what happens when a human carries out a computation.

Just how convincing was his argument that any such computation could just as well be carried out by a Turing machine? To buttress his case, Turing *proved* that a variety of complicated mathematical calculations could be done on Turing machines.* But the most audacious and far-reaching idea he came up with for testing the validity of what he had done was the *universal machine.*

Think of two natural numbers written on a Turing machine tape (in the usual decimal notation) separated by a blank square. The first number is to be the code of some Turing machine, and the second is to be an input to that machine:

CODE NUMBER OF A TURING MACHINE $\mathcal{M}$		INPUT TO $\mathcal{M}$

Now imagine a person given the task of working out what the Turing machine whose code number is the first number on the tape would do if confronted by the second number on the tape as input. The task is straightforward. The person could begin by obtaining the actual quintuples constituting the machine coded by the first number on the tape. Then, she could just "play" machine, that is she could simply do on the tape whatever the quintuples command.

Turing's analysis purported to demonstrate that any straightforward computational task can be carried out by a Turing machine. Applying this idea to the present task, one is led to imagine a Turing machine that, begun with the code number of a Turing machine $\mathcal{M}$ followed by a numerical input to $\mathcal{M}$ on its tape, would do exactly what the machine $\mathcal{M}$ would have done if confronted with that same input. *This would be one single Turing machine that alone could do anything that any Turing machine could do.* Turing tested this remarkable conclusion by setting himself the task of showing how one could actually produce the quintuples of such a "universal" machine. In a few pages of what now would be called code, he succeeded brilliantly in doing exactly this! [13]

*For example, Turing showed how to construct machines that could produce the sequences of 0s and 1s representing the binary representations of the real numbers e and π. He did the same for various other real numbers that come up in standard mathematics: roots of polynomial equations with integer coefficients and even the real zeros of Bessel functions.

People had been thinking about calculating machines since Leibniz's time and even earlier. Before Turing, the general supposition was that in dealing with such machines the three categories—machine, program, and data—were entirely separate entities. The machine was a physical object; today we would call it hardware. The program was the plan for doing a computation, perhaps embodied in punched cards or connections of cables in a plugboard. Finally, the data was the numerical input. Turing's universal machine showed that the distinctness of these three categories is an illusion. A Turing machine is initially envisioned as a machine with mechanical parts, *hardware*. But its code on the tape of the universal machine functions as a *program*, detailing the instructions to the universal machine needed for the appropriate computation to be carried out. Finally, the universal machine in its step-by-step actions sees the digits of a machine code as just more *data* to be worked on.

This fluidity among these three concepts is fundamental to contemporary computer practice. A *program* written in a modern programming language is *data* to the interpreter or compiler that manipulates it so that its instructions can actually be executed. In fact Turing's universal machine can be regarded as an interpreter, since it functions by *interpreting* successive quintuples in order to perform the tasks they specify.

Turing's analysis provided a new and profound insight into the ancient still of computing. The notion of computation came to be seen as embracing far more than arithmetic and algebraic calculations. And at the same time, the vision appeared of universal machines that in principle could compute everything that is computable. Turing's examples of specific machines are already instances of the art of programming; the universal machine in particular is the first example of an interpretative program. The universal machine also provides a model of a "stored program" computer in which the coded quintuples on the tape play the role of stored program, and in which the machine makes no fundamental distinction between program and data. Finally, the universal machine shows how hardware in the form of a set of quintuples thought of as the functioning of a mechanism can be replaced by equivalent software in the form of those same quintuples in coded form stored on the tape of a universal machine.

While working out his proof that there is no algorithmic solution to the *Entscheidungsproblem*, Turing did not suspect that similar conclusions were being reached on the other side of the Atlantic. Newman had already received a first draft of Turing's paper when an issue of the *American Journal of Mathematics* containing Alonzo Church's article "An Unsolvable Problem of Elementary Number Theory arrived in Cambridge." As was discussed in the previous chapter, Church had already shown in this article that there are algorithmically unsolvable problems. The article did not mention machines, but it did point to the concepts of λ-*definability* and *general recursiveness*,, both of which were proposed as explications of the intuitive notion of effective calculability. The two notions proved to be equivalent, and Church's unsolvable problem was in fact unsolvable with respect to either equivalent notion.

Although in this paper, Church had not drawn the conclusion that Hilbert's *Entscheidungsproblem* was unsolvable with respect to these notions, Volume 1 (1936), Number 1 of the new quarterly *Journal of Symbolic Logic* contained a brief note by Church in which he did exactly that. Turing quickly proved that his notion of computability was equivalent to λ-definability, and decided to spend some time in Princeton.

While much of what Turing had accomplished amounted to a rediscovery of what had already been done in the United States, his analysis of the notion of computation and his discovery of the universal computing machine were entirely novel, and went beyond anything that had been done in Princeton.* Kurt Gödel had been unconvinced by Church's proposals, and, writing many years later, he insisted that it was only Turing's analysis that finally convinced him of their correctness.[14]

Alan Turing in Princeton

When Alan arrived in Princeton on September 29, 1937, it was no longer what it had been over the past three years. Gödel had returned to Vienna, and Kleene and Rosser, having completed their studies, had gone their separate ways, launching their own careers. So, that left Alonzo Church. Although mathematicians in England did not typically bother to pursue doctorates, it was most convenient for Turing to arrange his stay at Princeton University by becoming a graduate student, really an anomalous status given his accomplishments. In the two years of his stay at Princeton, he completed a remarkable doctoral dissertation (with Alonzo Church as advisor). Since Gödel's proposition deemed undecidable in a given system could be seen to be true when viewed from outside the system, a natural approach was to add such a proposition to the given system as a new axiom, thus obtaining a new system in which that undecidable proposition was no longer undecidable. Of course, applying Gödel's methods, the new system would be seen to have undecidable propositions of its own. In his dissertation, Turing studied hierarchies of systems obtained by doing this over and over again.

---

*The first volume of the *Journal of Symbolic Logic* in which Church's proof of the unsolvability of the *Entscheidungsproblem* appeared also contained a short paper by the American logician E. L. Post that formulated a concept quite close to Turing's (Davis (1965, pp. 289–291)). Post was my teacher when I was an undergraduate at City College in New York City.

Another concept introduced in his dissertation is a Turing machine modified so that it could interrupt its computation to seek external information. By means of such machines, it becomes possible to speak of one of a pair of unsolvable problems being "more unsolvable" than the other. All in all, the ideas introduced in the dissertation were to provide the basis for the work of a succession of researchers.[15]

During his first year at Princeton, Alan had to make do with the meager stipend that his fellowship at Cambridge provided. This had been quite sufficient in Cambridge where room and board were also provided. However, during his second year he felt himself quite rich because he had been awarded the prestigious Procter Fellowship. Among the letters of recommendation written in support of his application for this fellowship was the following:

> June 1, 1937
>
> Sir,
>
> Mr A. M. Turing has informed me that he is applying for a Proctor [*sic*] Visiting Fellowship to Princeton University from Cambridge for the academic year 1937–1938. I should like to support his application and to inform you that I know Mr Turing very well from previous years: during the last term of 1935, when I was a visiting professor in Cambridge, and during 1936–1937, which year Mr Turing has spent in Princeton, I had opportunity to observe his scientific work. *He has done good work in branches of mathematics in which I am interested, namely: theory of almost periodic functions, and theory of continuous groups.* [emphasis added]
>
> I think that he is a most deserving candidate for the Proctor Fellowship, and I should be very glad if you should find it possible to award one to him.
>
> I am, Respectfully, John von Neumann[16]

Given that von Neumann had been deeply involved with Hilbert's program for the foundations of mathematics, it is very surprising that Turing's work on computability and his unsolvability proof for the *Entscheidungsproblem* are not mentioned in this letter. It is hard to believe that von Neumann didn't know about it.

I believe the key to making sense of this is the phrase "branches of mathematics in which I am interested." von Neumann, one of the great mathematicians of the century, an omnivorous reader with an almost photographic memory, evidently decided, after Gödel had demonstrated the futility of much of his work in this area, that he wanted nothing more to do with logic. He is even reputed to have said that after what Gödel did in

1931, he never again read a paper on logic.[17] This matter is of some importance because of the role of Turing's work on von Neumann's thinking about computers during and after World War II.

Some evidence is provided by a letter from von Neumann's friend and collaborator Stanislaw Ulam written to Turing's biographer Andrew Hodges.* This letter mentioned a game that von Neumann had proposed during the summer of 1938 when he and Ulam were traveling together in Europe; the game involved "writing down on a piece of paper as big a number as we could, defining it by a method which indeed has something to do with some schemata of Turing's." Ulam's letter also stated that "... von Neumann mentioned to me Turing's name several times in 1939 in conversations, concerning mechanical ways to develop formal mathematical systems." Ulam's letter makes clear that, whatever may have been the case earlier, by the outbreak of World War II in September 1939, von Neumann was well aware of Turing's work on computability.[18]

Turing's universal computer was a marvelous conceptual device that could alone execute any algorithmic task. But could one actually build such a thing? And aside from what such a machine could accomplish in principle, could it be designed and constructed so as to be able to solve real-world problems in an acceptable time frame, and using reasonable available resources? These questions were in Turing's mind from the very first. In an obituary article in the *Times* (of London), Turing's teacher Max Newman wrote:

> The description that he then gave of a "universal" computing machine was entirely theoretical in purpose, but Turing's strong interest in all kinds of practical experiment made him even then interested in the possibility of actually constructing a machine on these lines.[19]

Turing didn't confine himself to merely thinking about this possibility. To familiarize himself with the available technology, Turing went to the trouble of building a device using electro-mechanical relays that multiplied numbers written in binary notation. For this purpose he gained access to the Physics Department graduate student machine shop, and constructed various parts of the device, building the necessary relays himself.*

---

*Stanislaw Ulam (1909–1984) was a leading pure and applied mathematician who worked in many branches of mathematics and a good friend of von Neumann. One of his ideas ultimately led to an important way to extend the ordinary axioms of set theory in a manner that shed light on Gödel's work on the continuum hypothesis. Not everyone will applaud Ulam's most significant contribution: the basic design of fission-fusion thermonuclear weapons.

*Conveniently enough, the shop was in the Palmer Physics Laboratory located next door to Fine Hall, the mathematics building—there was even a passageway joining the two buildings.

Alan Turing's War

Turing returned to Cambridge in the summer of 1938. Although the war was still over a year in the future, he was recruited for an ongoing effort to break the codes used in German military communications. Codes and decoding had entered Turing's work and also Gödel's, but those codes were deliberately chosen to be transparent unlike the codes the Germans were using which were intended to be impenetrable. Indeed the Germans continued to believe throughout the war that their codes were impenetrable.

Following a pact between Nazi Germany and Communist Russia that surprised the world, German troops invaded Poland on September 1, 1939. Honoring a commitment, England and France declared war on Germany a few days later, and on September 4 Turing reported to Bletchley Park, a Victorian estate north of London, where a small team, mostly made up of academics, had gathered to read the messages the enemy intended to keep from them. The team was not destined to remain small. By the end of the war the estate was home to a number of "huts" in which various aspects of the decryption and analysis of messages was carried out. In addition to senior personnel, and of course the military, there were a considerable number of "wrens," women who had signed up for the Naval Auxiliary Corps and found themselves instead operating machines designed by Turing and his colleagues.

German military communications used a modified commercial encrypting machine called the *Enigma*. This machine had an alphabetic keyboard, and when a key for a particular letter was pressed, a letter would appear in a little window, the encrypted version of the original. When an entire message had been encrypted, it would be sent out by ordinary radio telegraphy. The intended recipient would enter the encrypted letters into another Enigma machine, and the original message would appear. Inside the machine were a number of rotating wheels acting to change the match between input letter and the letter's encrypted version from letter to letter. In the military version, security was enhanced by an additional plugboard. Each day there would be a different initial setting of the machine which had to be the same for the sender and the recipient.

A group of Polish mathematicians had done an amazing job of deciphering German Enigma messages before the war began, but when the Germans added a layer of complexity to the machines, they were stymied, and passed their work onto the British. The cryptanalysts at Bletchley Park liked to work on puzzles, and at times they were deeply engrossed in the intellectual aspects of their problems, and were enjoying themselves. But the work was deadly serious.

Turing's particular responsibility was the communications between German submarines and their home base. Ships bringing badly needed supplies to the British Islands were being destroyed by these submarines at an alarming rate. If the U-boats weren't stopped, it seemed entirely possible that England would simply be starved out.

Success in decrypting the Enigma traffic was helped by a seized code book from a captured submarine and by some carelessness on the part of senders that unintentionally gave away crucial information. But the crucial role was played by Turing who saw how to design a machine (called a "Bombe" for no reason anyone seems to be able to recall) that proved very effective in using this information to deduce the settings of the German's Enigma on any given day.

Fittingly enough, the Bombes systematically carried out chains of logical reasoning that eliminated one possible Enigma configuration after another from among the huge number possible, until only a few were left. These were then worked over by hand until the correct one emerged. Unlike the usual experience with an untried gadget, Turing's Bombes, built from his design, worked correctly as soon as they were made.*

At Bletchley Park, Turing was affectionately called "the prof" and his eccentricities became the sources of anecdotes. Years later people spoke of his habit of keeping his tea mug chained to the radiator. Perhaps the most revealing anecdote from the Bletchley Park days concerns how Turing learned to shoot a rifle. In the dark days of 1940 and 1941 when England seemed open to invasion, the Churchill government formed a citizen's militia, the "Home Guard." Although, because of the importance of his work, he wasn't required to join the Home Guard, Alan Turing decided to join so he could learn to shoot a rifle.

Recruits for the Home Guard were required to attend regular drills, and, after a while, Turing decided that these were a waste of time—so he stopped attending. Called to order by one Colonel Fillingham with a reputation for easily becoming "apoplectic," Turing patiently explained that he had joined only to learn to shoot, and now that he had become an excellent shot, he no longer had any reason to attend. Said the colonel: "But it is not up to you whether you attend ... it is your duty as a soldier to attend. ... You are under military law." The colonel reminded Turing that in applying to join, he had filled out a form with the question: "Do you understand that by enrolling in the Home Guard you place yourself liable to military law?" To which Turing replied that he had indeed answered that question, but that the answer he had written was "No." In considering that question, it

*Gordon Welchman, a mathematician who was six years older than Turing, added a very important feature to Turing's design that greatly enhanced its performance. Readers interested in the technical details of how the Enigma traffic was deciphered are referred to Welchman's account (Welchman, 1982) and to Hodges (1983). Hinsley and Stripp (1933) contains interesting accounts of life in Bletchley Park during the war by a number of the participants in the deciphering effort.

was evident to Turing that there would be no advantage to him in a "yes" answer.[20]

In addition to being amusing, this anecdote reveals much of Turing's character. He tended to ignore much of the social framework in terms of which most of us act, and, in any situation, he would think things through, starting from scratch, seeking the optimal action. Most people confronting a question like the one on the Home Guard application form would realize that only an affirmative answer would be acceptable, but Turing took the question at face value, and thought seriously about what would be the best answer. Although this way of thinking worked very well for Turing in his scientific research, it did not work so well in his interactions with people and social institutions, and ultimately, years later, it led to disaster.

Turing became friendly with Joan Clarke, a young mathematician enlisted in the Bletchley Park endeavor. He found himself in love with her, proposed marriage and was gladly accepted. She found it definitely worrisome, when, a few days later, he told her of his homosexual "tendencies," but she was willing to carry on with the engagement. A few months later, shortly after they had taken a vacation trip together, Turing decided that although he really loved Joan, it just wouldn't work, and he broke off the engagement. Apparently this was the first and last time that he permitted himself to imagine an amorous relationship with a woman.

Meanwhile, Turing never stopped thinking about the applicability a universal machine. He guessed that this notion of universality held the secret of the enormous power of the human brain, that in some manner our brains are actually universal machines. He imagined that if a universal machine could be built, it could be made to play games like chess, that it could be induced to learn much as a child does, and ultimately it could be made to exhibit behavior one would be led to call "intelligent." There was much conversation along these lines in Bletchley Park, and Turing even sketched algorithms that a machine could use in playing chess. At the same time, some of the hardware needed for building a universal machine was being developed right there in Bletchley Park.

Some of the messages intercepted in England, communications that originated at the highest levels of the Nazi regime, were not Enigma-encrypted, and were not transmitted by ordinary telegraphy. It was soon realized that that they had the characteristics of teleprinter output. This was a system in which each individual letter in a text was represented by a row of holes in a paper tape. Unlike the older Morse code telegraphy, no human operator was required. It seemed that the Germans were using a single machine that could encrypt and transmit a message as a single operation. The recipient would have a machine that would do the decoding. At Bletchley Park, this system was called "fish," and Turing's teacher, Max Newman, undertook the task of deciphering it. Some of the methods to be used were playfully called *turingismus* indicating their source.* But turingismus required the processing of lots of data, and for the decryption be of any use, the processing had to be done very quickly.[21]

In the 1930s most people in the United States and Europe had radios in their homes. In those days, before transistors had been invented, radios contained a number of vacuum tubes (called "valves" in Britain). In use, these glowed like low-intensity light bulbs and became quite hot. Like light bulbs, they burned out frequently and had to be replaced. When one's radio stopped working, one could pull the tubes from their sockets and bring them into a shop for testing. After replacing the ones that had gone bad, the radio would usually come back to life. The RCA catalog of tubes listing hundreds of different models of tubes with their specific characteristics, was indispensable to engineers and popular with hobbyists.

In March 1943, Alan Turing was returning from a visit of several months in the United States where he had helped launch the American effort to construct their own Bombes and to take over the monitoring of naval Enigma traffic. He whiled away the time during his Atlantic passage by studying this catalog, for it had been found that vacuum tubes could be used to carry out the kind of logical switching for which electric relays had previously been used. And the tubes were fast: their electrons moved at speeds close to that of light, while relays depended on mechanical motion.

Vacuum-tube circuits had been used experimentally for telephone switching, and Turing had made contact with a gifted engineer, T. Flowers, who was at the forefront of this research. Under the direction of Flowers and Newman, a machine that was essentially a physical embodiment of turingismus was rapidly developed. This machine, dubbed the *Colossus*, was an engineering marvel, containing 1500 vacuum tubes. It was the world's first electronic automatic calculation device. Not surprisingly, the computations it carried out were logical rather than arithmetic in nature.

Intercepted German communications in the form of punched paper tapers were fed to the machine by an extremely fast tape reader: as the tape moved through the reader, beams of light passing through the holes in the paper were intercepted by photoelectric cells which passed the signal on to the Colossus. It was important that the tape be read rapidly in order not to slow down the operation of the vacuum tube circuits. Flowers's outstanding feat was not only getting an operational machine constructed in a few months, but also managing to get useful work done by a machine containing so many tubes. Indeed many had thought that the inevitable frequency of tube failures would make this impossible.

---

* *ismus* is a German suffix used much like the English *ism*.

By the time the war ended in 1945, Turing had become convinced that vacuum tube circuits could be used to construct a universal computer. He had developed a practical knowledge of vacuum tube electronics and had devoted considerable thought to practical issues of implementation. He had thought about the great variety of situations and problems to which such a machine could be applied. He needed only the support and facilities to bring the great project to fruition.

CHAPTER 8

Making the First Universal Computers

Who Invented the Computer?

Modern computers are complex amalgams of logic and engineering and it would be possible to single out any one person as *the* inventor. Nevertheless in 1973, in resolving a patent dispute (in the case *Honeywell v. Sperry Rand*), a judge came close to doing just that. As our story moves from the underlying logical ideas behind modern all-purpose computers to their construction, engineering issues and the people who were able to deal effectively with them come to the fore. Accounts of the history of computing have made varying claims, and before continuing our story, it's worth having a quick look at the cast of characters.

JOSEPH-MARIE JACQUARD (1752–1834). The Jacquard loom that weaves cloth with a pattern specified by a stack of punched cards revolutionized weaving practice, first in France, and eventually all over the world. With perhaps understandable hyperbole, it is commonly said among professional weavers that this was the first computer. Although it is a wonderful invention, the Jacquard loom was no more a computer than is a player piano. Like a player piano, it permits a mechanical device to be controlled automatically by the presence or absence of punched holes in an input medium.[1]

CHARLES BABBAGE (1792–1871). Babbage proposed to use punched cards like Jacquard's for his never-built analytical engine. He owned a self-portrait of Jacquard in the form of a weaving.

ADA LOVELACE (1815–1852). Her father, Lord Byron, never saw her after her first year. She had a great passion for mathematics and was particularly passionate about Babbage's proposed analytical engine. She translated a French memoir about the analytical engine to which, with Babbage's encouragement, she added extensive comments. She has been called the world's first computer programmer—a major programming language has been named Ada in her honor. Her aphorism relating the analytical engine to Jacquard's loom is often quoted:

> We may say most aptly that the Analytical Engine weaves *algebraical patterns* just as the Jacquard-loom weaves flowers and leaves.[2]

CLAUDE SHANNON (1916–2001) In his master's thesis at MIT (published in 1938), Shannon showed how George Boole's algebra of logic could be used to design complex switching circuits. This thesis "...helped to change digital circuit design from an art to a science."[3] His mathematical theory of information has played a crucial role in contemporary communication technology. Shannon did pioneering work in computer algorithms for chess playing. He showed how to construct a universal Turing machine with only two states. (Shannon was my boss in 1953 when I had a summer job at Bell Labs.)

HOWARD AIKEN (1900–1973). His *Automatic Sequence Controlled Calculator*, constructed by IBM for Harvard University using electric relays and inaugurated in 1944, did everything Babbage envisioned. Having developed a machine specifically intended for the kind of number crunching needed by physicists and engineers, Aiken found it difficult to see that a machine intended to be all-purpose could be effective for this kind of computation.

KONRAD ZUSE (1910–1995). He was a German computer pioneer who, working in total isolation during World War II with very limited support from the Nazi government, managed to design and construct a working calculator which like Aiken's used electric relays. However, unlike Aiken's machine, Zuse's used binary rather than decimal arithmetic, and he took advantage of the simplified construction this enabled.

JOHN ATANASOFF (1903–1995). This obscure physicist at Iowa State University (working with his assistant Clifford Berry) designed and built a small special-purpose calculator based on vacuum tube electronics during the years leading to the U.S. entry into World War II. Although this machine could only deal with problems of a very special kind, it was important because it demonstrated the usefulness of vacuum tube circuits for computation.[4]

JOHN MAUCHLY (1907–1980). Although Mauchly was trained as a physicist, it was his vision that was behind the development of the world's first large scale number-crunching electronic calculator known as the ENIAC at the Moore School of Electrical Engineering of the University of Pennsylvania in Philadelphia. Mauchly had visited Atanasoff in Ames, Iowa, and apparently had been given full access to his electronic calculator.

J. PRESPER ECKERT JR. (1919–1995). It was the brilliant electrical engineer Eckert whose remarkable efforts were mainly responsible for the successful construction of the ENIAC.

HERMAN GOLDSTINE (1913–2004). The mathematician Herman Goldstine, inducted into the U.S. Army in 1942, was assigned to the Ballistic Research Laboratory of Army Ordnance as a first lieutenant. As the Army's representative on the ENIAC project, he brought von Neumann into the group at the Moore School. In the later disputes with Eckert and Mauchly, he supported von Neumann. After the war, he became von Neumann's chief collaborator in work concerning computation. His book on the history of computation (Goldstine, 1972) emphasized von Neumann's role, and was criticized for that reason. (In 1954, he was the person to whom I had to apply for permission to use the computer at the Institute for Advanced Study.)

EARL R. LARSON (1911–2001). He was the U.S. District judge who, in 1973, found the patent that Eckert and Mauchly had obtained on the ENIAC invalid. His opinion included the statement:

> Eckert and Mauchly did not themselves first invent the automatic electronic digital computer, but instead derived that subject matter from one Dr. John Vincent Atanasoff.[5]

John von Neumann and the Moore School

As we saw in the previous chapter, John von Neumann had taken on the task of explaining Hilbert's program at the symposium on the foundations of mathematics in Königsberg in 1930. This was the symposium at which Kurt Gödel dropped his bombshell asserting that he had established the necessary incompleteness of formal systems for mathematics, and it was apparently von Neumann who was the first at that conference to grasp the significance of what had been accomplished.

Soon after that, von Neumann wrote Gödel quite excitedly: " ...I achieved a result that seems to me to be remarkable. For I was able to show that the consistency of mathematics is unprovable." What von Neumann had seen was that by using Gödel's methods, it could be proved that systems like those Hilbert had in mind were inadequate to prove their own consistency. As we have already noted, by the time Gödel received this letter, he had come to the same conclusion and sent by return mail a printed abstract containing that result.

John von Neumann was a vain, brilliant man. He was used to putting his stamp on a mathematical subject by the sheer force of his powerful intellect. He had devoted considerable effort to the problem of the consistency of arithmetic, and in his presentation at the Königsberg symposium, had even come forward as an advocate for Hilbert's program. Seeing at once the profound implications of Gödel's achievement, he had taken it one step further—proving the unprovability of consistency, only to find that Gödel had anticipated him. It was enough.

Although he was full of admiration for Gödel, and even lectured on his work, he refused to have anything more to do with logic. He is said to have boasted that after Gödel, he simply never read another paper on logic. Logic humiliated him, and von Neumann was not used to being humiliated. Nevertheless, von Neumann's need for powerful computational machinery eventually forced him to return to logic.

As with Turing, von Neumann's wartime work called for large-scale computation. But, where the cryptanalytic work at Bletchley Park emphasized the side of computation involving symbolic patterns, so in tune with Turing's earlier work, it was old-fashioned heavy number crunching that von Neumann needed. Not surprisingly, he eagerly accepted an opportunity to participate in an exciting project at the Moore School of Electrical Engineering in Philadelphia to construct a powerful electronic calculator, the ENIAC. It was the 30-year-old mathematician Herman Goldstine who brought von Neumann into the ENIAC project. As Goldstine tells the story, von Neumann learned of the ENIAC project quite fortuitously when the two met for the first time at a railway station during the summer of 1944. Von Neumann soon joined the discussions with the ENIAC group at the Moore School.

The Colossus with its 1,500 vacuum tubes was already an engineering marvel; the ENIAC with 18,000 tubes was simply astonishing. The conventional wisdom of the time was that no such assemblage could do reliable work; it was believed that a tube would be bound to fail every few seconds. The chief engineer on the ENIAC project, John Presper Eckert, Jr., was largely responsible for the project's success.

Eckert insisted on very high standards of component reliability. Tubes were operated at extremely conservative power levels, and the failure rate was kept to three tubes per week. The ENIAC was an enormous machine, occupying a large room, and was programmed by connecting cables to a plugboard rather like an old-fashioned telephone switchboard.[6] The ENIAC was modeled on the most successful machines available at the time for dealing with the kinds of problems expected to be posed to the ENIAC: differential analyzers.

Differential analyzers were not "digital" devices operating on numbers digit by digit. Rather numbers were represented by physical quantities that could be measured (like electric currents or voltages) and components were linked together to emulate the desired mathematical operations. These "analog" machines were limited in their accuracy by that of the instruments used for the measurements. The ENIAC was a digital device, the first electronic machine able to deal with the same kind of mathematical problems as differential analyzers. Its designers built it of components functionally similar to those in differential analyzers, relying on the capabilities of vacuum tube electronics for greater speed and accuracy.[7]

By the time that von Neumann began meeting with the Moore School group, it was clear that there were no important obstacles to the successful completion of the ENIAC, and attention was focused on the next computer to be built, tentatively called the EDVAC. Von Neumann immediately involved himself with the problems of the *logical* organization of the new machine. As Goldstine recalled:

> Eckert was delighted that von Neumann was so keenly interested in the logical problems surrounding the new idea, and these meetings were scenes of greatest intellectual activity.
>
> This work on the logical plan for the new machine was exactly to von Neumann's liking and precisely where his previous work on formal logics came to play a decisive role. Prior to his appearance on the scene, the group at the Moore School concentrated primarily on the *technological* problems, which were very great; after his arrival he took over leadership on the *logical* problems.[8]

In June 1945 John von Neumann produced his famous *First Draft of a Report on the EDVAC* which, in effect, proposed that the soon-to-be-built EDVAC be realized as a physical model of Turing's universal machine. Like the tape on that abstract device, the EDVAC was to possess a storage capability, called "memory," holding both data and coded instructions. In the interest of practicality, the EDVAC was to have an arithmetic component that could perform each basic operation (addition, subtraction, multiplication, or division) in a single step, whereas in Turing's original conception, these operations would need to be built up in terms of primitive operations such as "move one square to the left."

Whereas the ENIAC had performed its arithmetic operations on numbers represented in terms of the ten decimal digits, the EDVAC was to enjoy the simplicity made possible by binary notation. The EDVAC was also to contain a a component exercising logical control by bringing instructions to be executed one at a time from the memory into the arithmetic component. This way to organize a computer has come to be known as the "von Neumann architecture," and although what computers are made of today is very different from the parts that were available for the EDVAC, today's computers are for the most part still organized according to this basic plan.[9]

The EDVAC report never advanced beyond the draft stage, and it is incomplete in a number of ways. In particular, there are many places where a reference to be inserted later is indicated. Turing's name is never mentioned, but his influence is evident to the discerning eye. The notion that the EDVAC should be "all purpose" is mentioned more than once. Like Turing, von Neumann surmised that some of the remarkable capability of the human brain was the result of its possessing the power of a universal computer. In the EDVAC report, von Neumann refers over and over again to the analogy between the brain and the machine he is discussing. He notes that vacuum tube circuits can be designed to behave in many ways like the neurons in our brains, and, without worrying about the engineering details, he describes how the arithmetic and logical control components needed for the EDVAC could be built of such circuits.

Although the report is almost entirely devoid of references, there is a conspicuous exception: there are a number of references to a paper by a pair of MIT researchers, published in 1943, that set out a mathematical theory of such idealized neurons. One of the authors of this paper later stated that he had been directly inspired by Turing's 1936 article (the one in which his universal machine was explicated), and in fact the paper has just one reference—to Turing's article. More revealing still, the authors take the trouble to demonstrate that a universal Turing machine can be modeled using their idealized neurons, and cite this fact as the principal reason for believing that their work is on the right track.[10]

Eckert and Mauchly came to bitterly protest von Neumann's release of the EDVAC Report under his own name. An element of controversy, which will probably never be fully resolved, is the question of how much of the EDVAC report represented von Neumann's contribution. Although Eckert and Mauchly later denied that von Neumann contributed very much, shortly after the report appeared they wrote as follows:

> During the latter part of 1944, and continuing to the present time, Dr. John von Neumann . . . has fortunately been available for consultation. He has contributed to many discussions on the logical controls of the EDVAC, has prepared certain instruction codes, and has tested these proposed systems by writing out the coded instructions

> for specific problems. Dr. von Neumann has also written a preliminary report in which most of the results of earlier discussions are summarized. ... In his report, the physical structures and devices ... are replaced by idealized elements to avoid raising engineering problems which might distract attention from the logical considerations under discussion.[11]

There is other evidence that von Neumann wanted to be sure that the machine he was specifying was as close as was practically possible to being universal. So he emphasized the "logical control" of a computer as being crucial for its being "as nearly as possible *all purpose*."[12] In order to test the general applicability of the EDVAC, von Neumann wrote his first serious program, not for the kind of number-crunching application for which the machine was mainly developed, but rather to simply sort data efficiently. The success of this program helped to convince von Neumann that "... it is legitimate to conclude already on the basis of the now available evidence, that the EDVAC is very nearly an 'all purpose' machine, and that the present principles for the logical controls are sound."[13]

Articles written within a year of the EDVAC report confirm von Neumann's awareness of the basis in logic for the principles underlying the design of electronic computers. The introduction to one such article states:

> In this article we attempt to discuss [large scale computing] machines from the viewpoint not only of the mathematician but also of the engineer and the logician, i.e., of the ... person or group of persons really fitted to plan scientific tools.[14]

Another article clearly alludes to Turing's ideas even as it emphasizes that purely logical considerations are not enough:

> It is easy to see by formal-logical methods that there exist codes that are *in abstracto* adequate to control and cause the execution of any sequence of operations which are individually available in the machine and which are, in their entirety, conceivable by the problem planner. The really decisive considerations from the present point of view, in selecting a code, are of a more practical nature: simplicity of the equipment demanded by the code, and the clarity of its application to the actually important problems together with the speed of its handling those problems. It would take us much too far afield to discuss these questions at all generally or from first principles.[15]

It is well understood that the computers developed after World War II differed in a fundamental way from earlier automatic calculators. But the nature of the difference has been less well understood. These postwar machines were designed to be all-purpose universal devices capable of carrying out any symbolic process whatever, so long as the steps in the process were specified precisely. Of course, some processes may require more memory than is available or may simply take too long to be feasible, so these machines can only be approximations to Turing's idealized universal machine. Nevertheless it was crucial that they had a large "memory" (corresponding to Turing's infinite tape) in which instructions and data could coexist.

This fluid boundary between what was instruction and what was data meant that programs could be developed that treated other programs as data. In early years, programmers mainly used this freedom to produce programs that could and did modify themselves. In today's world of operating systems and hierarchies of programming languages, the way has been opened to far more sophisticated applications. To an operating system, the programs that it launches (e.g., word processor or email program) are data for it to manipulate, providing each program with its own part of the memory and (when "multitasking") keeping track of the tasks each needs carried out. Compilers translate programs written in one of today's popular programming languages into the underlying instructions that can be directly executed by the computer: for the compiler, these programs are data.

After the experience with the ENIAC and with the Colossus, those interested in computational equipment would not settle for speeds of operation slower than what they knew could be obtained using vacuum tube electronics. For an all-purpose computer modeled on Turing's universal machine, a physical device was needed that could function as an appropriate large memory.

On the tape of Turing's abstract universal machine, moving from a particular square to another distant one required a laborious process of repeatedly moving one square at a time. This was fine for Turing's purposes in 1936. Those theoretical "machines" were not meant to do anything practical. However, a fast electronic computer needed a fast memory. This required that the data stored at any place in the memory should be directly accessible in a single step, that is, the memory should be "random access."

In the late 1940s, two devices offered themselves as candidates for use as computer memory: the mercury delay line and the cathode ray tube. The delay line consisted of a tube of liquid mercury; data was stored in the form of an acoustic wave in the mercury bouncing back and forth from one end of the tube to the other. Cathode ray tubes are familiar in early TV sets and computer monitors. Data could be stored as a pattern on the surface of the tube. There were serious engineering problems with both of these devices, but fortunately for the EDVAC project, Eckert had developed improved delay lines during the war for use with radar. However, by the early 1950s cathode ray tubes had become the preferred memory medium.

In discussions of this period, the new computers that were being developed are usually referred to as embodying "the stored program concept" because, for the first time, the programs to be executed were stored within the computer. Unfortunately this terminology has served to obscure the fact that what was really revolutionary about these machines was their universal all-purpose character, while the stored-program aspect was only a means to an end. The point of view of Turing and von Neumann is conceptually so simple and has so much become part of our intellectual climate, that it is difficult to understand how radically new it was. It is far easier to appreciate the importance of a new invention like the mercury delay line, than of a new and abstract idea.

Eckert later claimed that he had already thought of the so-called "stored program concept" well before von Neumann had appeared on the scene. His evidence was a memo that spoke of "automatic programming" set up on "alloy discs" or "etched discs." There is nothing here that even remotely suggests the concept of the all-purpose computer with a large flexible memory in which instructions and data cohabit. But to characterize the great advance that had been made as the "stored-program concept" is to invite such confusion.[16]

The bitterness between Eckert and Mauchly on the one hand, and von Neumann and Goldstine on the other came to a head when Eckert and Mauchly attempted to develop a commercial product based on their work. They sought patents for the ENIAC and for the EDVAC. Their application for an EDVAC patent got nowhere precisely because the circulation of von Neumann's draft report had placed it in the public domain. As already explained, they did receive a patent for the ENIAC, later found invalid by a court. Eckert and Mauchly were certainly prescient in envisioning the commercial possibilities for all-purpose electronic computers, but they were unable to profit from their prophetic insight.[17]

With the departure of Eckert and Mauchly, the Moore School lost much of its edge, and von Neumann and Goldstine went on to develop a computer at the Institute for Advanced Study in Princeton using a cathode ray tube memory. A special purpose tube developed by RCA Corporation on which von Neumann had set his hopes did not work out, but the English engineer Frederic Williams (1911–1977) developed methods by which ordinary cathode ray tubes could be used effectively as computer memories and for some years the "Williams memory" dominated the scene. A number of machines similar to the Institute machine were built, affectionately termed "johnniacs" after Johnny von Neumann. When IBM decided that it was time to market all-purpose electronic computers, their first model (the 701) was quite similar to the johnniacs.*

Alan Turing's ACE

At the end of World War II, Britain's National Physics Laboratory (NPL) underwent a considerable expansion including a new Mathematics Division. J. R. Womersley (1907–1958), appointed head of this division, had seen the practical implications of Turing's 1936 *Computable Numbers* paper quite early on. In 1938 he had gone so far as to undertake the design of a universal machine using electric relays, but abandoned the idea because he saw that such a device would be too slow. On a visit to the United States in February 1945, he saw the ENIAC and obtained a copy of von Neumann's EDVAC report. His reaction was to hire Alan Turing.

By the end of 1945, Turing had produced his remarkable ACE (Automatic Computing Engine) report. One detailed comparison of the ACE report with von Neumann's EDVAC report, notes that whereas the latter "is a draft and is unfinished ... more important ... is incomplete ..." the ACE report "is a complete description of a computer, right down to the logical circuit diagrams" and even including "a cost estimate of £11,200." In a list of ten problems that might be handled by the ACE, Turing, showing the breadth of his vision, included two that did not directly involve numerical data: playing chess and solving simple jigsaw puzzles.[18]

Turing's ACE was a very different kind of machine from von Neumann's EDVAC, corresponding closely to the different attitudes of the two mathematicians. Although von Neumann was concerned that his machine be truly "all-purpose," his emphasis was on numerical calculation and the logical organization of the EDVAC (and of the later johnniacs) was intended to expedite this direction. Since Turing saw the ACE being used for many tasks for which heavy arithmetic was inappropriate, the ACE was organized in a much more minimal way, closer to the Turing machines of the *Computable Numbers* paper.

Arithmetic operations were to be carried out by programming: by software rather than hardware. For this reason, the ACE design provided special mechanism for incorporating previously programmed operations in a longer program.[19] Turing was particularly caustic concerning a proposal to modify the ACE in a von Neumann direction:

*My personal introduction to computer programming occurred in the spring of 1951 when I began writing code for the ORDVAC, a johnniac built at the University of Illinois in Champaign-Urbana. In the summer of 1954, I wrote a program (not unrelated to Leibniz's dream) that ran on the original johnniac at the Institute for Advanced Study. That computer can be seen at the Smithsonian Institution in Washington.

> [It] is . . . very contrary to the line of development here, and much more in the American tradition of solving one's difficulties by means of much equipment rather than by thought. . . . Furthermore certain operations which we regard as more fundamental than addition and multiplication have been omitted.[20]

Turing's minimalist ideas were destined to have little or no influence on computer development. But in retrospect one can see that so-called *microprogramming* which makes the the most basic "basement-level" computer operations directly available to the programmer was anticipated by the ACE design. Also, the personal computers we use nowadays are built around silicon microprocessors that are in effect universal computers on chips, and these have become more elaborate. An opposing paradigm, the so-called RISC (reduced instruction set computing) architecture, adopted by a number of computer manufacturers, uses a minimal instruction set on the chip, with needed functionality supplied by programming, again very much in line with the ACE philosophy.

On February 20, 1947, Turing addressed the London Mathematical Society on the subject of the ACE in particular and digital electronic computers in general. He began by referring to his 1936 *Computable Numbers* paper:

> I considered a type of machine which had a central mechanism, and an infinite memory which was contained on an infinite tape. . . . One of my conclusions was that the idea of a 'rule of thumb' process and a 'machine process' were synonymous . . . Machines such as the ACE may be regarded as practical versions of . . . the type of machine I was considering . . . There is at least a very close analogy . . . digital computing machines such as the ACE . . . are in fact practical versions of the universal machine.[21]

Turing went on to raise the question of ". . . how far it is in principle possible for a computing machine to simulate human activities." This led him to propose the possibility of a computing machine programmed to learn and permitted to make mistakes. "There are several theorems which say almost exactly that . . . if a machine is expected to be infallible, it cannot also be intelligent . . . But these theorems say nothing about how much intelligence may be displayed if a machine makes no pretence at infallibility."

This was an oblique reference to Gödel's incompleteness theorem about which there will be more to say in the next chapter. Turing concluded his lecture with a plea for "fair play for computers" that should not be expected to be more infallible than human beings, and a suggestion that chess playing would be an appropriate exercise on which to begin. All of this was at a time when not a single one of these devices had yet been completed! By all reports, the audience was stunned into silence.[22]

When the Bletchley Park leaders were having trouble getting adequate resources and support, they sent a letter to Winston Churchill who immediately saw to it that they got what they needed. Construction of the ACE could command no such priority, and, in addition, the administration of the NPL behaved in a most inept manner.

T. Flowers, who had done such a bravura job of getting the Colossus built, would have been the ideal person to build the ACE. Although he did some work on delay lines for computer memory under contract with NPL, he was much too busy with postwar telecommunications work to be of much help.

There was concern about the minimalist design of the ACE, perhaps tinged with a feeling that the Americans were the ones to trust with technological issues rather than an eccentric English don. What this don had done to help win the war remained a deeply guarded secret for many years. When Williams showed that his cathode ray tube memory would work, he was offered a contract to work on the ACE, which he declined. This negotiation was quite inept on the part of the NPL administration. They imagined that Williams could be hired to build the NPL computer, whereas Williams had access to sufficient resources to build his own computer at Manchester.

Finally Turing had had enough and left, first taking up a fellowship at Cambridge, and then accepting a job offer from the University of Manchester where his old friend and war-time comrade Max Newman was starting a computer project. Afterwards with a change of personnel, a small version of the ACE was built successfully at NPL. Called the "Pilot ACE" it worked well for years.

Eckert, von Neumann, and Turing

As historians well know, the way a story is told changes with time, often drastically. In the story of what is usually called the "stored program concept," there have been three principal versions. The first account saw the concept as the product of von Neumann's genius as promulgated in his EDVAC report. Eckert cried "foul" and insisted that he had proposed a stored-program computer before von Neumann had joined the Moore School group. The EDVAC report, he asserted, represented the joint thinking of the group. Publications appeared supporting Eckert's position.[23] Turing's name was not mentioned at all. Supporting von Neumann's claim and oblivious to Turing's role, Goldstine wrote:

> von Neumann was the first person, as far as I am concerned, who understood explicitly that a computer essentially performed logical functions, and that the electrical aspects were ancillary.[24]

Of course, Turing understood that very well indeed.

The gap between the thinking that went into the ENIAC and the universal computer is so immense that I find it difficult to believe that Eckert had envisioned anything like the latter. When Turing complained about "the American tradition of solving one's difficulties by means of much equipment rather than by thought," he likely had the ENIAC in mind. From Turing's conclusion that "the idea of a 'rule of thumb' process and a 'machine process' [are] synonymous" it is plain that converting numbers from decimal to binary and back is the most trivial of machine operations.

Not seeing this, and concerned with the need for quantities to be input and output in decimal notation, Eckert and Mauchly solved their problem by designing their behemoth of a machine that carried out all of its internal operations in decimal notation. Many problems that occur in practice require finding approximate values for certain limit operations of the calculus. Because the analog machines, called differential analyzers, included special modules that could compute such approximations, Eckert and Mauchly incorporated modules performing similar functions in their ENIAC. But this is totally unnecessary and inappropriate for a digital machine. Calculus textbooks describe methods for calculating these values requiring nothing more than the four basic operations of arithmetic.

Eckert did perform one immense service in connection with the EDVAC and that was to propose the mercury delay line as an answer to the problem of the need for a large memory. Eckert had worked with these delay lines for use with radar and knew a great deal about them. Therefore, it is telling that in the memo he later cited as proving that he had thought of the "stored program concept" first, he spoke of "automatic programming" set up on "alloy discs" without mentioning the delay lines that he knew all about and that would have been far more creditable as a memory medium.

It is interesting to contrast von Neumann's view of computer programming as an activity with Turing's; von Neumann made it clear that he thought of it as a clerical task requiring little intellect. A revealing anecdote tells of a practice at the Institute for Advanced Study computer facility of using students to translate computer instructions written using human-readable mnemonics into machine language by hand.

A young hot-shot programmer proposed to write an assembler that would do this conversion automatically. Von Neumann is said to have responded angrily that it would be wasteful to use a valuable scientific tool to do a mere clerical job. In his ACE report, Turing said that the process of computer programming "should be very fascinating. There need be no real danger of it ever becoming a drudge, for any processes that are quite mechanical may be turned over to the machine itself."[25]

Although the Eckert and the von Neumann versions of the story are still heard, a third version has become prominent. This third version has von Neumann getting the idea of a practical universal computer from Turing's work. In 1987, when I wrote an article expounding that point of view, I felt myself to be very much alone.[26] Since then information about Turing's role in decrypting German communications during the war has become much more widely known. Also many people have become aware of the shameful way he was persecuted for having had a homosexual affair.

Breaking the Code, a successful play performed in London and on Broadway that was also the basis for a television play shown on PBS has dramatized these matters as well as the importance of Turing's mathematical ideas.[27] Television documentaries have also told his story. And so, lo and behold, Alan Turing's name was on the list of the 20 greatest "scientists and thinkers" of the twentieth century proposed by *TIME* magazine (in its March 29, 1999 issue).* Said *TIME*:

> So many ideas and technological advances converged to create the modern computer that it is foolhardy to give one person the credit for inventing it. But the fact remains that everyone who taps at a keyboard, opening a spreadsheet or a word-processing program, is working on an incarnation of a Turing machine.

Exactly! And here is what *Time* had to say about von Neumann:

> Virtually all computers today from $10 million supercomputers to the tiny chips that power cell phones and Furbies, have one thing in common: they are all "Von Neumann machines," variations on the basic computer architecture that John von Neumann, building on the work of Alan Turing, laid out in the 1940s.

A Grateful Nation Rewards Its Hero

When Turing arrived in Manchester in the fall of 1948, it was still recovering from the war, and there were neighborhoods that retained their grim aspect left over from the city's role in the early days of the Industrial Revolution. One writer uses a famous book by Friedrich Engels as a source in commenting on the squalor of working-class housing in the Manchester of 1844:

> What he [Engels] ... describes ... fall[s] within a uniform context of mass immiseration, degradation, brutalization, and imhumanization, the like of which had never before been seen on the face of the earth. ... On reaching these courts, he finds himself met with an assault of "dirt and revolting filth, the like of which is not to be found ... [and] without qualification the most horrible dwellings I have until now beheld ... In one of these courts, right at the entrance ... is a

*Kurt Gödel was another of the 20.

privy without a door. The privy is so dirty that the inhabitants can only enter or leave by wading through puddles of stale urine and excrement."[28]

Of course, mass sanitation had seen dramatic improvements during the ensuing century, and in any case, someone of Turing's social class would not have lived in a working-class neighborhood. Nevertheless, Turing's association with a member of the "lower" classes was to lead to disaster.

One can only imagine how bitter Turing must have felt about the inept management at NPL that had squandered his talent and had nullified the confident dream he had revealed in his ACE Report and in his address to the London Mathematical Society. Meanwhile, computers were being built. At Cambridge University, Maurice Wilkes (1913–2010) directed the construction of an EDVAC-type computer called the EDSAC. Unlike Turing's situation at NPL, Wilkes had adequate funding in house for his project. It must have been particularly galling to Turing to recollect that at NPL, he had scorned a memo from Wilkes as being in "the American tradition of solving one's difficulties by means of much equipment rather than by thought."

By 1949 the EDSAC was operational and open for business. The supposed discoveries by Wilkes and his collaborators of microprogramming and the systematic use of subroutines, both of which were clearly spelled out in Turing's ACE report, can only have added to his distress. At Manchester, where Turing was supposed to be somehow directing the computer project there, Williams made it quite clear that he was not interested in some mathematician's ideas about the construction of his computer. The Mark I Manchester computer, also running successfully in 1949, was a brilliant vindication of Williams' technique for using "off the shelf" cathode ray tubes as his memory devices, soon copied in American computers. But again, its basic logical design derived from von Neumann's EDVAC report and not from Alan Turing.[29]

About Turing's ACE, Herman Goldstine remarks that although the design was "attractive in some respects," it "did not in the long run flourish and selection weeded it out."[30] The suggestion that this was somehow the result of a kind of natural selection is really unfair. The Pilot ACE embodying Turing's ideas worked perfectly well. There is no reason to think that a full scale ACE-style computer would not have worked well if the organization and resources to build one had been there.

The issue is best understood in the more general context of the question of which computer functions should be supplied by hardware and which by software. Turing had proposed a relatively simple machine in which a lot was left to be supplied by software, but where, in compensation, the programmer had very substantial control of underlying machine operations.

This would be particularly advantageous for writing programs that are intended to carry out logical rather than numerical calculations. As the field developed people continued to debate this tradeoff, for example in connection with RISC (reduced instruction set computing) architecture.*

When Turing arrived at the University of Manchester in 1948, few people had any notion of what he had done during the war, although he continued to be consulted by the government. He had been hired with the understanding that he would exercise some administrative functions in connection with Williams' Mark I computer, but as things worked out, the engineers pretty much ran their own show, and what Turing did along these lines was carried out in a rather desultory fashion. Instead of using his position to introduce some of the elegant ideas proposed in his ACE report to make the programmer's job pleasant and easy, he became a user of the machine, and worked directly with the 0s and 1s of machine language. He worked on some computational problems that he had thought about before the war, but his interests soon turned to biology.

He sought to answer the question of how living things, starting out as assemblages of identical cells, managed to develop the varied forms encountered in the natural world. This problem of *morphogenesis* gave rise to differential equations, and Turing naturally turned to the computer for information about the solutions of these equations. While using the machine for the kind of number-crunching application he had proposed to go beyond in popular articles and public addresses, he demonstrated his continuing imaginative vision of the potential of computers for human-like intelligence.

It was just before Christmas 1951 that Turing managed to launch a brief affair with the 19-year-old youth, Arnold Murray. Murray was a very bright young man from a poor working-class family. When Turing accosted him in the street, he was on probation, having been caught in a petty theft. Turing invited him to his house which must have seemed a palace to Murray. Less than a month after Christmas, Turing returned home one evening to discover that his house had been broken into and burglarized. Although the total value of what had been taken amounted to no more than £50, Turing was quite upset. It turned out that Murray had a pretty good idea who had carried out the theft, namely someone he knew named Harry. Harry had evidently felt secure in robbing a homosexual who presumably

*I personally wrestled with the basically number-crunching instruction set of von Neumann's Institute for Advanced Study computer during the summer of 1954. I was implementing an algorithm for testing the truth of sentences of PA (defined in the Appendix to Chapter 6) that involved addition, but not multiplication. (The editors of an anthology of technical papers in this field of computational logic said in their preface, referring to my program: "In 1954 a computer program produced what appears to be the first computer generated mathematical proof" (Siekmann and Wrightson, 1983, p. ix).) I don't doubt that the ACE instruction set would have been a good deal more suitable for my purpose.

would not dare go to the police. He was certainly right that a prudent man in Turing's position would not do anything so foolish as to go to the police. But that is exactly what Turing did.

The police had little trouble working out what had happened between Turing and Murray, and when confronted, Turing denied nothing. He did not believe that there was anything shameful or wrong about the nature of his sexual feelings or in the harmless ways he went about fulfilling them. Be that as it may, the law was quite clear on the matter: what Turing and Murray had done in giving one another pleasure were acts of "gross indecency," punishable by up to two years in prison.

The judge before whom Turing's case came, acting out of what he believed were humane motives, permitted Turing to escape prison if he would agree to be treated by hormone injections for a year in order to diminish his sex drive. The hormone chosen was estrogen, and, whatever effect it may have had on Turing's sex drive, it had the incidental effect of causing him to grow breasts.

In October 1938, Turing saw Walt Disney's *Snow White and the Seven Dwarfs*. "He was very taken with the scene where the Wicked Witch dangled an apple on a string into a boiling brew of poison, muttering

> Dip the apple in the brew.
> Let the Sleeping Death seep through.

It seems that he took pleasure in chanting this verse over and over again.[31] On June 7, 1954, Alan Turing ended his life by eating an apple slice that had been dipped into a cyanide solution. There has been much speculation about what led him to this irreversible act. The play *Breaking the Code* proposes that governmental authorities were objecting to the vacation trips abroad that, after his conviction, had become his most promising source of sex partners. Sex in England had become dangerous, perhaps too dangerous to attempt. That the authorities in the atmosphere of the 1950s, did object to his trips abroad, seems not in the least implausible. After his conviction, he lost his security clearance. But there was no way to erase the secret information he carried in his brain. What is definitely known is that a man he had met on a trip to Norway had been stopped by the police and deported when he came to England to visit Turing. Alas, it seems all too possible that Alan Turing was hounded to his death by the governing authorities of a nation whose unsung savior he had been.

CHAPTER 9

Beyond Leibniz's Dream

In his address before the London Mathematical Society, Turing said:

> I expect that digital computing machines will eventually stimulate a considerable interest in symbolic logic ... The language in which one communicates with these machines ... forms a sort of symbolic logic.[1]

The connection between logic and computation to which Turing alludes has been a principal theme of this book. Nevertheless, readers may still ask: how is it that logic and computation are related? What does arithmetic have to do with reasoning? A clue is provided by a colloquial use of the verb "to reckon," in which it does not have its usual meaning: "to calculate."

> I *reckon* he's sweet talking her in the moonlight right now.

We are listening to the melancholy hero of a grade-B film speaking of his rival, not knowing (as we do) that it was our hero who had already won her heart. In statement, he is not thinking of arithmetic; he is talking about *reasoning*. He reasons based on what he thinks he knows about his rival's perfidious ways. The connection between calculation and reasoning suggested by this use of the word "reckon" is genuine and profound. Reckoning with numbers is a form of reasoning, and a great deal of the reasoning that people do can be regarded as a kind of computation. It is very interesting that, as our example shows, this relationship is generally understood, at least on a subliminal level. We can see this appreciation as well when someone is described as a calculating sort of person.

Reducing logical reasoning to formal rules is an endeavor going back to Aristotle. It was the underlying basis for Leibniz's dream of a universal computational language. And it underlay Turing's achievement in showing that all computation could be carried out on his universal machines. Computation and logical reasoning are indeed two sides of the same coin. This insight is used not only to make it possible to program computers to perform a bewildering variety of tasks, but indeed in the very way that computers are designed and built.

In January 1999, the American Association for the Advancement of Science held its annual meeting in Anaheim, California under the heading

Challenges for a New Century. This was a serious, very rich meeting with lectures and symposia covering an enormous array of topics. Surveying the program, a symposium entitled *On the Unusual Effectiveness of Logic in Computer Science* caught my attention. The presentations by five leading computer scientists (who are also logicians) attested to the correctness of Turing's prediction. Their statement in the program about the symposium reads in part:

> ...during the past twenty-five years there has been an extensive, continuous, and growing interaction between logic and computer science. In many respects, logic provides computer science with both a unifying foundational framework and a tool for modeling ...This symposium provides an overview of the surprising effectiveness of logic in computer science by presenting some of the areas in which logic played a crucial role in computer science: artificial intelligence, computational complexity, distributed computing, database systems, and design verification.[2]

Eliza, Watson, and Deep Blue

In 1965, Joseph Weizenbaum, a computer scientist at MIT, wrote a program called ELIZA that engaged in a dialogue with a user entering English sentences from a keyboard.[3] ELIZA was meant to simulate in very crude manner a psychotherapist of the non-directive school. Here is a transcription of my actual conversation with a version of ELIZA:

ELIZA: HOW DO YOU DO: PLEASE TELL ME YOUR PROBLEM.

MARTIN: My problem is indescribable.

ELIZA: WHY DO YOU SAY YOUR PROBLEM IS INDESCRIBABLE?

MARTIN: Because whenever I attempt to describe it, I cry.

ELIZA: IS THAT THE REAL REASON?

MARTIN: Perhaps not.

ELIZA: YOU DON'T SEEM QUITE CERTAIN.

MARTIN: Ah, certainty.

ELIZA: CAN YOU EXPAND ON THAT?

MARTIN: Yes. But it will lead to intimate matters.

ELIZA: WHY ARE YOU SO POSITIVE?

MARTIN: I try to maintain my dignity.

ELIZA: YOUR DIGNITY?

MARTIN: You wouldn't understand that.

ELIZA: WE WERE DISCUSSING YOU—NOT ME.

MARTIN: I have had quite enough. Goodbye!

ELIZA: YOU SAY YOU HAVE HAD QUITE ENOUGH.

MARTIN: I said Goodbye.

ELIZA: CAN YOU ELABORATE ON THAT?

MARTIN: Goodbye.

ELIZA: GOODBYE. THAT WILL BE $75, PLEASE.

Of course ELIZA understands nothing. The program implements a very simple algorithm that responds to key words when they are present and otherwise echoes the user's sentence. For example "you" is a key word and its presence will generally cause the response: WE WERE DISCUSSING YOU—NOT ME. The sentence "You are my mother" contains the two key words "mother" and "you." It will elicit the response TELL ME MORE ABOUT YOUR FAMILY because the key word "mother" overrides the key word "you." The key word "yes" elicits the response WHY ARE YOU SO POSITIVE? Notice that ELIZA responds to "Goodbye" only when it begins a sentence.

There has been an almost unbelievable advance in computer technology from the vacuum tube machines of Turing's time (including those I personally wrote code for in the 1950s), to the solid state IBM 7090 that Weizenbaum would already have had available when he wrote the ELIZA program, and from that to the "smart phones" many of us carry in pockets or purses. But all of them embody the same underlying logic that Turing proposed. From his abstract universal machine with no limitations of space or time, Turing inferred that to the extent that a physical device was able to carry out a small number of basic instructions sufficiently rapidly and was equipped with sufficient data storage, it would be able to carry out any algorithm in a reasonable time. My own Android phone can do a remarkably good job of transcribing my spoken English into text, and I can download to it an app (the current term for a computer program) that can play a much better chess game than I can.

In 1950, Alan Turing published his classic essay "Computing Machinery and Intelligence" in which he proposed a way to discuss the question of whether a computer could be said to exhibit intelligent behavior without getting into the morass of philosophical and theological questions into which that question often leads. For this purpose he proposed an objective easy-to-administer test that he called the *imitation game*: if a computer can be programmed to carry on a conversation with a reasonably intelligent person on whatever topics are raised so effectively that a user cannot tell whether he or she is talking to a person or to a machine, then said Turing, we should be prepared to agree that the computer is exhibiting intelligence.[4] He wrote:

> I believe that in about fifty years' time it will be possible to programme computers ... to play the imitation game so well that an average interrogator will have not more than a 70 per cent chance of making the right identification after five minutes of questioning. ... I believe that at the end of the century the use of words and generally educated opinion will have altered so much that one will be able to speak of machines thinking without expecting to be contradicted.[5]

Regarding this last point, later in this chapter we will encounter scholars who resolutely refused to agree that activities that we regard as involving "knowing" and "thinking" when people do them, should have the same words applied to these activities when machines do them, even when they do them quite well.

In 2011, there was considerable excitement when a behemoth of a computer named Watson developed by IBM, succeeded in defeating the best human players in the popular television quiz program *Jeopardy*. Watson did not have access to the Internet, but its massive database included the entire *Wikipedia* encyclopedia and much else beside. The *Jeopardy* format helped Watson seem to be more fluent in ordinary English than it actually was. The players were furnished with a "clue" and were to come up with the question to which that clue was the answer. For example, presented with the clue "he succeeded in turning Aristotle's syllogistic reasoning into manipulation of equations," the contestant should reply: "Who was George Boole?" The structured format required of the contestant was a real advantage for Watson's programmers who did not have to cope with the kind of free-wheeling dialog that the Turing test requires.

Watson used the words and phrases making up the clue to search its data base for matches. It used an algorithm to provide a numerical score for each of the possible responses its search suggested, and only signified that it was ready to respond if and when a sufficiently high score was achieved. The producers of *Jeopardy* took advantage of the opportunity to display their showmanship. Watson was furnished with a stage presence in the form of a loud speaker, and its response was enunciated by a speech synthesizer. But quite apart from this hullabaloo, Watson's performance was a spectacular achievement by the IBM researchers.

While computational linguists continue to seek the holy grail of imbuing computers with the capability of using ordinary language, it is natural to seek machine intelligence in domains not dependent on ordinary language. One such domain is the game of chess. It would be difficult to deny that a person playing even a reasonably good game of chess is exercising intelligent thought. And it is common knowledge that chess-playing programs that play very good games of chess are readily available. Most ordinary players must set these to play at less than the programs' best in order not to be regularly defeated.

In February 1996, the chess-playing computer Deep Blue managed to defeat world champion Garry Kasparov. May we then say that Deep Blue exhibited intelligence? In an article written in his usual provocative style, the philosopher John R. Searle tells us that Deep Blue cannot properly even be said to play chess.

> Here is what is going on inside Deep Blue. The computer has a bunch of meaningless symbols that the programmers use to represent the positions of the pieces on the board. It has a bunch of equally meaningless symbols that the programmers use to represent options for possible moves. The computer does not know that the symbols represent chess pieces and chess moves, because it does not know anything.[6]

To hammer the point home, Searle has recourse to a variant of a parable that he has made quite famous. The original story tells of a man in a room who receives symbols from outside the room and by looking things up in a book determines which symbols he should send out in reply. It turns out that the book is so written that the symbols flying back and forth constitute a conversation in Chinese. But the man knows no Chinese and has no idea what the symbols represent. Leaving aside what conclusion one may draw from this bizarre tale, let us move on to Searle's "Chess Room":

> Imagine that a man who does not know how to play chess is locked inside a room, and there he is given a set of, to him, meaningless symbols. Unknown to him, these represent positions on a chessboard. He looks up in a book what he is supposed to do, and he passes back more meaningless symbols. We can suppose that if the rule book, i.e., the program, is skillfully written, he will win chess games. People outside the room will say, "This man understands chess, and in fact he is a good chess player because he wins." They will be totally mistaken. The man understands nothing of chess, he is just a computer. And the point of the parable is this: if the man does not

> understand chess on the basis of running the chess-playing program, neither does any other computer solely on that basis.

Readers of this book will perhaps notice the arbitrary separation of software from hardware in this example. The man in the room simply functions as a crude version of a universal computer. Of course a bare-bones computer doesn't play chess. It is only the man together with the instruction book for which any such claim might be made. Here is my version of Searle's parable:

> A precocious child whose mother is passionate about chess becomes tired of watching her play and demands that he be allowed to play her opponent. His mother agrees on the condition that he move the pieces only when she tells him to and exactly where she says. He does as requested and doing what his mother whispers in his ear achieves a checkmate. Observing the scene, Searle tells us that the child doesn't know anything about chess, and is certainly not playing chess. Who could disagree?

It is part of the contemporary philosophers' method to tell stories they know to be quite preposterous for the purpose of bringing out connections that might otherwise not be apparent. But it may not be entirely pointless to bring the Chess Room down to earth. I once had a colleague who had been part of the team that designed Deep Thought, the powerful chess-playing computer that was the predecessor of Deep Blue. He provided me with some numbers on the basis of which I calculated that if the hardware and software constituting Deep Thought were put in the form of a book (more likely a library) of instructions that a human being could carry out, then it would take several years to do the processing needed to make one move. Better put a family in that Chess Room, so the children can take over when the parents die! Otherwise, no game will actually be completed.

Searle tells us that Deep Blue "has a bunch of meaningless symbols." Well, if you could look inside Deep Blue while it was in operation, you wouldn't see any symbols, meaningful or not. At the level of circuits, electrons are moving around. Just as, if you look inside Kasparov's skull while he is playing, you wouldn't see any chess pieces, you'd see neurons firing. The way our brains are organized to deal with what we think of as symbolic information is still only dimly understood. The way computers (like Deep Blue) are organized for this purpose is much better understood, because the engineers and programmers build it in. But in both cases, processes that function at something like the molecular level are integrated into patterns that we can think of as involving symbolic manipulation. Searle tells us that the symbols that Deep Blue has are meaningless. Well, whatever does a pawn or a knight "mean"? This is not a useful question.

Searle makes much of the fact that Deep Blue doesn't "know" that it is playing chess. In fact he insists it doesn't "know" anything. Actually, professional knowledge engineers would likely insist that Deep Blue does know all sorts of things. For example, it "knows" to which squares a bishop at a given square can move. It all depends on what "knows" means. Be that as it may, we can agree that Deep Blue does not know that it is playing chess. Can we therefore conclude that it is not in fact playing chess? Here's another parable:

> Anthropologists studying the Xlupu people of northern New Guinea have made a remarkable discovery of something which must surely be one of the greatest coincidences of all time. Although the Xlupu have lived in total isolation until this year, it appears they engage in a religious ceremony in which pairs of them engage in a symbolic ritual exactly equivalent to our game of chess. They do not use a board or pieces, but rather make intricate designs in boxes of sand. It is only because Dr. Splendid, the leader of the anthropological expedition that first encountered the Xlupu, is himself an enthusiastic amateur chess player, that he was able to see in the patterns being drawn, equivalents of the successive moves in a chess game.

Are these Xlupu playing chess? They surely don't *know* that that is what they are doing. Ah! Searle might reply (yes, I am putting words in his mouth): "But the Xlupu are conscious, and Deep Blue is not." The question of whether a programmed computer might ever be conscious has played a major role in discussions of these matters by Searle and others. Whatever may come to pass in the future, one certainly must agree that Deep Blue is not conscious.

Our consciousness is a principal way in which each of us experiences his or her unique individuality. But we know it only from the inside. We experience our own consciousness but not that of anyone else. I experience my consciousness as an internal conversation. My wife assures me that her consciousness is dominated by visual images. Are her consciousness and mine really the same kind of thing? What is it and what purpose does it serve? As I write, I seek the right word, and (when I'm lucky) it appears in my consciousness from the depths below. How my brain manages to do such a clever thing I have no idea. The simple truth is that at this time the phenomenon of consciousness remains mysterious.

Computers Play Board Games

People talked about computers playing chess before there were any computers. It was discussed at Bletchley Park during the code breakers' leisure time. Claude Shannon published a paper explaining how it could be done in 1950 when the existing machines were far too weak for the purpose. A player of a game like chess is faced at each stage of play with a choice of possible moves.

Programs that play such games are traditionally conceived in terms of data arranged in a structure called a "tree." At each stage the various moves that a player can make are thought of as forming "branches." The *length* of such a tree can show all the moves of a possible game from beginning to end. The *width* of the tree indicates the range of possible moves at a given stage of play. If we call the players "white" and "black" with white having the first move, we can imagine a branch for each possible white move. From the end of each such branch, we can imagine branches emanating, corresponding to each possible black move. Continuing in this way, the tree quickly becomes enormous. Designers of computer programs to play the game seek to prune the tree to make it manageable.

Shannon suggested that one should provide a way to represent the values of given positions of pieces positioned on a chessboard so that an appropriate move could maximize the value of the position that would result from that move. For chess, players traditionally value a pawn at 1, knights and bishops at 3, a rook at 5, and the queen at 9. Using these numbers together with the estimated value of such features as control of the center of the board, mobility of the pieces, and safety of the king, a numerical value can be obtained for a given position.

A good player tries to look ahead for several moves, anticipating the opponent's likely response to a given move. Once the computer becomes sufficiently powerful to support such a look-ahead for a number of moves, this simple idea can form the skeleton of an effective chess-playing program. The most powerful chess playing programs available at this time, able to defeat the best human players, augment this with a look-up library of opening moves and responses as well as of end-game scenarios in addition to various other features that add subtlety to the game-playing algorithm.

Until very recently, mastery of the ancient Chinese game of Go was securely in the hands of human players. On the face of it, Go is much more complex than chess. Where chess is played on a board with 64 squares on which the pieces can be placed, Go is played on a 19×19 board with 361 places where pieces can be placed.

As in chess the winner is entirely determined by the moves the players make, taking alternate turns and choosing as they wish among the legal moves in the position they face. Like checkers, but unlike chess, in Go there is just one kind of piece. In Go the pieces are called stones and are traditionally colored black or white. A computer program to play such a game is based on an algorithm that "prunes" the game tree by choosing favorable moves at each stage of play. Go was such a formidable problem, at least in part, because its tree is so very long and extremely wide. It was thus a remarkable achievement when the program AlphaGo, developed by the DeepMind group, defeated the South Korean Go master player Lee Sedol in four games of a five-game match played in March 2016. DeepMind was founded in London in 2010 and acquired by Google in 2014. It has grown from a group of a dozen or so computer scientists and software engineers to a team of more than 500 people working on artificial intelligence and machine learning. Susan Dickey, a Google software engineer, was able to arrange an hour-long meeting with one of these computer scientists, Thore Graepel, via video connection with Susan and me in a Google office in San Francisco and Thore in London.*

As early as the 1940s, researchers explored mathematical models suggested by biological nervous systems.[7] These models are called *neural networks.* They are conceptualized as networks of interconnected so-called "neurons." Each neuron has several input channels that receive numerical signals from other neurons and one output channel that transmits such signals. Each input channel has a number associated with it called its *weight*; at each stage the total signal received by a neuron is the number obtained by multiplying each input value to a channel by the weight for that channel and then adding all of those numbers.

What if anything the neuron transmits at that stage is decided depending on the value of this total signal. Although one may imagine a neural network implemented as a huge number of tiny electronic gadgets interconnected by a network of wires, in practice a neural network has only a virtual existence as a program stored in a computer. As computers have become more powerful, the variety and complexity of neural networks that have been built, studied, and applied to serious problems has increased dramatically.

Contemporary neural networks are usually imagined as consisting of interconnected layers and have the capability of changing some of the weights of their individual neurons. This capability enables such neural networks to *learn* how to perform various computational tasks. They are provided with examples of what is wanted and use this data to to modify themselves so as to steadily improve their performance. This technique for *training* a neural network is often referred to as *deep learning.*

Such deep learning techniques have been very successful in training neural networks to identify specific objects in a video image. Video images exist in the form of an array of pixels stored in a computer. Neural networks have

*My friend Susan Dickey and I were colleagues in the Computer Science Department at NYU many years ago. I am grateful to her for arranging this meeting and for the excellent notes she took. I am also very grateful to Thore for the clear and careful explanations he provided. This discussion is based on what Thore told us and on Susan's notes. Of course any errors are mine.

been trained to distinguish a bird from a squirrel and to identify individual human faces. The networks that accomplish this feat are usually designed to be "convolutional": what people and animals accomplish by scanning a visual field through eye movements, a convolutional neural network accomplishes by carrying out the enormous number of computations needed to recognize that objects in different parts of a video image are the same kind of thing.

These computations are greatly facilitated by having special "graphical processing unit" chips as part of the hardware. These GPU chips were originally designed and manufactured for machines intended for playing computer games.[8] The algorithm implemented in AlphaGo can be thought of as a tree search algorithm taking advice from two auxiliary convolutional neural networks: a *policy* network and a *value* network. Both networks were trained by giving them access to a huge library of games played between expert players over many years. The policy network estimates how likely various possible moves are to to be played by an expert Go player stating from a particular arrangement of pieces on the Go board.

Thore told us, "This policy network was already a pretty good Go player. I am a pretty good amateur player myself. On my first day at DeepMind, David Silver invited me to play against this network, and I was sure I would win. But it beat me! Then I was sure I wanted to be part of the AlphaGo team."

The other network, the value network, starting with any Go position, estimates for one of the players the probability that that player will win. Convolution plays a role similar to that in object recognition: just as a bird in one corner of a video image needs to be recognized as still a bird when it's in a different part of the image, a certain configuration of the stones in one part of the Go board needs to be recognized as similar to one in a different part. The hardware used by AlphaGo includes GPUs to make this work efficiently.

In 2016 AlphaGo was brought to Korea to challenge Lee Sedol to a five-game match. Thore said, "Although we had great confidence in AlphaGo, you never know what chance will come up with in a tournament. Fortunately, AlphaGo won the first three games. By the fourth game, our team was actually rooting for Lee Sedol, and was happy for him when he succeeded." Then AlphaGo won the fifth game.

In May 2017 at the Future of Go Summit in China, AlphaGo won all three games against world champion Ke Jie. Thore continued, "Go players have started to pick up patterns of play that AlphaGo invented. In training for the May competition, we let AlphaGo play against itself and create new games of higher quality, resulting in a set of training data for a stronger version of AlphaGo. With this kind of bootstrapping, a machine learning system can continue to grow."

Computers, Brains, and Minds

Turing and von Neumann were both led to compare computers with the human brain for an excellent reason. Knowing that people were capable of so many diverse patterns of thought, they conjectured that we can do so many very different things because embedded in our brain is a universal computer. That's the reason that von Neumann was so struck by a theory of artificial neurons when he set out to design the EDVAC. What universal computers can do is to execute algorithms. Searle says, "... in fact humans do rather little that is literally computing. Very little of our time is spent working out algorithms ..." Is he so sure?

Answer the question: have you ever read anything by Charles Dickens? The answer (yes or no) comes welling up from the depths. How do we do it? We have no idea. But the hypothesis that it is done by some kind of algorithmic processes that access the required information from some databases in our brain is on the face of it quite attractive.

Research on computer processing of raw visual data entering a computer from one or more TV cameras is very suggestive of the kind of process needed to produce the sharp picture our brain presents to us from the raw data going from the retina to the brain. We don't *know* that the way we do such things is by means of our brains carrying out algorithms, but we certainly don't know that that's not how it's done.

Roger Penrose is an outstanding mathematician and mathematical physicist who has done exciting work on the geometry of the universe. He has considered the question of whether the functioning of the human mind is fundamentally algorithmic, and has invoked Gödel's incompleteness theorem to answer the question with a resounding "No." One way to express Gödel's theorem is as follows:

> *Given an algorithm that produces true statements about the natural numbers one-after-another, we can always obtain another true statement about the natural numbers, let us call it the Gödel sentence, that is not generated by that algorithm.*[9]

Penrose argues that no particular algorithm proposed to be equivalent to the mind's working can possibly be adequate for that purpose because by an act of "insight," we can see that the Gödel sentence for that algorithm is true. This argument is deeply fallacious for a reason that Turing explained in his lecture to the London Mathematical Society in 1947, four decades before Penrose wrote on the subject.

What Turing pointed out is that Gödel's theorem applies only to algorithms that generate only true sentences. But no human mathematician can claim infallibility. We all make mistakes! So there is nothing in

Gödel's theorem to preclude the mathematical powers of a human mind being equivalent to an algorithmic process that produces false as well as true statements.[10]

Searle and Penrose reject the conjecture that the human mind is in all essentials equivalent to a computer. But both of them tacitly accept the premise that whatever the human mind may be, it is produced by the human brain, subject to the the laws of physics and chemistry. Kurt Gödel, on the other hand, was quite prepared to believe that the brain is in effect a computer, but rejected the idea that there is no mind beyond what the human brain can do. Most readers will recognize that the classical "mind-body" problem is at the core of Gödel's concerns. His position that the mind is in some way independent of our existence as physical entities is usually called "Cartesian dualism."[11]

This discussion has taken us far beyond Leibniz's dream, to a realm somewhere between philosophy and science fiction. However, taking note of what has become of computers since the days of the EDVAC and ACE reports, should not lead us to forget the continued viability of Turing's vision of a universal computer. This was present in abstract form in his 1936 *Computable Numbers* article, and laid out more explicitly in the ACE report: Given sufficient fast memory storage, and sufficiently fast data processing, one universal machine can execute any algorithmic process. One would indeed be well advised to be cautious in predicting what computers may or may not be able to do in the future.

Epilogue

We have followed the lives of a group of brilliant innovators spanning three centuries. All of them in one way or another were concerned with the nature of human reason. Their individual contributions added up to the intellectual matrix out of which emerged the all-purpose digital computer. Except for Turing, none of them had any idea that his work might be so applied.

Leibniz saw far, but not that far. Boole could hardly have imagined that his algebra of logic would be used to design complex electric circuits. Frege would have been amazed to find equivalents of his logical rules incorporated into computer programs for carrying out deductions. Cantor certainly never anticipated the ramifications of his diagonal method. Hilbert's program to secure the foundations of mathematics was pointed in a very different direction. And Gödel, living his life of the mind, hardly thought of applications to mechanical devices.

This story underscores the power of ideas and the futility of predicting where they will lead. The Dukes of Hanover thought they knew what Leibniz should be doing with his time: working on their family history. Too often today, those who provide scientists with the resources necessary for their lives and work try to steer them in directions deemed most likely to provide quick results. This is not only likely to be futile in the short run, but more important, by discouraging investigations with no obvious immediate payoff, it short-changes the future.

Notes

Preface to Second Edition

[1]For a recent comprehensive treatise on embedded computers, see Fisher et al. (2005). For my article on the history emphasizing Turing's role, see Davis (1988). For the suggested influence of my work see Leavitt (2006, p. 6). (All references are to the bibliography at the back of the book.)

Introduction

[1]The quotation is from Ceruzzi (1983, p. 43). Howard Aiken (1900–1973) founded the Harvard Computation Laboratory and was instrumental in the design and construction of large-scale calculating devices at Harvard during the 1940s and early 1950s.

[2]The quotation is from an address to the London Mathematical Society (Turing, 1992, pp. 112); (Copeland, 2004, p. 383). Alan Turing is the subject of Chapters 7 and 8 of this book.

Chapter 1

[1]For biographical information about Leibniz, I have relied mainly on Aiton (1985).

[2]For Leibniz's *Dissertatio de Arte Combinatoria* (alas in the original Latin), see Leibniz (1858/1962).

[3]Leibniz's mathematical work in Paris is discussed in Aiton (1985) and more extensively in Hofmann (1974).

[4]Quoted from Leibniz (1685/1929).

[5]For Leibniz's writing about machinery for reasoning and for equation solving, see Couturat (1961, p. 115).

[6]Readers interested in the mathematical details of the development of the calculus by Newton and Leibniz and their predecessors will enjoy the fine treatment in Edwards (1979). The reader is also referred to Bourbaki (1969, pp. 207–249) for an excellent account of the historical development of the calculus.

[7]There is another interesting story (but one that really belongs in another book) about Leibniz's differential and integral calculus: his systematic use of "infinitesimal" numbers. Infinitesimals were supposed to be positive numbers so very tiny that no matter how many times such a number is added to itself, the number 1 (or even the number .0000001) will never be reached. The legitimacy

of such quantities was challenged from the outset; the philosopher Bishop Berkeley scoffed at infinitesimals as "ghosts of departed quantities." By the end of the nineteenth century, mathematicians were in agreement that the use of infinitesimals could not be justified (although physicists and engineers continued to employ them). Discussion of infinitesimal methods as used by Leibniz as well as their eventual rehabilitation in the twentieth century by the logician Abraham Robinson will be found in the book (Edwards, 1979) already cited. The *Scientific American* article (Davis and Hersh, 1972) gives another account of Robinson's achievement.

[8]Aiton (1985, p. 53).

[9]Mates (1986, p. 27). See also pp. 26–27 of this source for more about about these remarkable women, for information about Leibniz's beliefs about the intellectual capabilities of women, and for further references.

[10]The letter to L'Hopital quoted was dated April 28, 1693 (Couturat, 1961, p. 83). The quote from Couterat is from the same page of the same source. For the "thread of Ariadne" see Bourbaki (1969, p. 16).

[11]The letter from Leibniz to Jean Galloys Leibniz (1849/1962) on his Universal Characteristic was dated December 1678. The translation from French is mine.

[12]Gerhardt (1978, vol. 7, p. 200).

[13]Parkinson (1966, p. 105).

[14]For Leibniz's logical calculus of which a small "sample" is exhibited here, see Lewis (1918/1960, pp. 297–305). Leibniz did not use the $=$ symbol, but instead used ∞. The interesting article (Swoyer, 1994) gives a thorough reconstruction of this system from a twentieth century perspective.

[15]For some discussion of Leibniz's attempts to go beyond Aristotle's analysis, see Mates (1986, pp. 178–183).

[16]Huber (1951, pp. 267–269).

[17]Aiton (1985, p. 212).

Chapter 2

[1]Information about Leibniz's friendship with Princess Caroline and his correspondence with Samuel Clarke is from Aiton (1985, pp. 232, 341–346) and the articles "Caroline (1683–1737)" and "Clarke, Samuel (1675–1729)" *in Britanica* (1910/11).

[2]Biographical information about George Boole is principally from MacHale (1985).

[3]MacHale (1985, pp. 17–19).

[4]"Gross appetites and passions" (MacHale, 1985, p. 19).

[5]MacHale (1985, pp. 30–31).

[6]MacHale (1985, pp. 24–25).

[7]MacHale (1985, p. 41).

[8]Among the most important of the laws of algebra are the *commutative* laws for addition and multiplication:

$$x + y = y + x, \qquad xy = yx,$$

and the *distributive* law

$$x(y + z) = xy + xz.$$

We are using the usual algebraic convention of writing, for example, xy instead of $x \times y$.

[9]Multiplication of two differential operators (which is taken to mean applying first one and then the other) doesn't always obey the commutative law.

[10]Boole's gold medal (MacHale, 1985, pp. 59–62, 64–66). In addition to Boole's work employing the methods of the calculus, he published a paper in two parts in the *Cambridge Mathematical Journal* for 1842 that can be thought of as founding a new and important branch of algebra, the theory of invariants. However, after this first contribution, Boole never again worked on invariants. We will be considering invariants again in the chapter on David Hilbert.

[11]Boole's casual attitude to proof in connection with limit processes may be contrasted with contemporary efforts on the continent to develop an appropriate rigorous foundation for such matters. Interested readers are referred to Edwards (1979), especially Chapter 11.

[12]The Scottish philosopher Sir William Hamilton is not to be confused with his contemporary, the Scottish mathematician, Sir William Rowan Hamilton.

[13]Boole (1854, pp. 28–29).

[14]Daly (1996); Kinealy (1996)

[15]MacHale (1985, p. 173).

[16]MacHale (1985, p. 92).

[17]MacHale (1985, p. 107).

[18]MacHale (1985, pp. 240–243).

[19]MacHale (1985, p. 111).

[20]MacHale (1985, pp. 252–276).

[21]The modern notation for the intersection of x and y is $x \cap y$ rather than xy. Also the empty set is usually represented by the Danish letter $\emptyset$ rather than by 0. Of course the notation he used was important for Boole because it made it easy to connect with ordinary algebra.

[22]Boole restricted the operation $+$ to classes having no elements in common. Here we follow contemporary usage and do not enforce this restriction. So $x + y$ is the class of things belonging to x or y or both. Nowadays one speaks of the *union* of x and y, written $x \cup y$. Also, Boole restricted the notation $x - y$ to the case that the class that y represents is part of the class that x represents. But there is no need for this restriction either.

[23]Boole (1854, p. 49).

[24]As Boole emphasizes, what is involved algebraically in demonstrating the validity of a syllogism is the *elimination* of one variable from two simultaneous equations in three variables.

Although Boole realized perfectly well that propositions of the form "All X are Y" could be represented in his algebra as $X(1-Y)=0$, he preferred to use $X=vY$ where v is what he called an *indefinite symbol.* This was apparently suggested by the mathematician Charles Graves (MacHale, 1985, p. 70). It was really a terrible idea and a quite unnecessary complication of Boole's system.

[25]Boole's method of relating secondary propositions to his algebra of classes was to bring *time* into the picture. With each proposition Boole would in effect associate the *class* of instants of time for which that proposition was true. To say that proposition X is true, Boole would write $X=1$ meaning that the class of instants in which the proposition is true encompasses the entire time span under consideration. Likewise, $X=0$ would express that X is false, because there are no instants of time in which X is true. Given a proposition $X\&Y$ which expresses the truth of both X and Y, the set of instants in which it is true is just the set intersection XY. Finally, for a proposition *if* X *then* Y to be true, what is required is that any time that X is true, Y is also true, that is that there is no time when X is true and Y is false. As an equation: $X(1-Y)=0$. (Boole, 1854, pp. 162–164).

[26]Boole (1854, pp. 188–211).

Chapter 3

[1]For Russell's letter, Frege's reply and Russell's later comment, see van Heijenoort (1967) pp. 124–128.

[2]For Frege's notorious diary as well as Michael Dummett's comment, see Frege (1996).

[3]I am very much indebted to Professor Lothar Kreiser of the University of Leipzig who graciously replied to my request for information about Frege. See his masterful biography (Kreiser, 2001). Terrall Bynum's brief biography in Bynum (1972) was also helpful.

[4]I found Craig (1978) an excellent source on German history. For the origins of the First World War, see also Geiss (1967); Kagan (1995). A number of postcards from Frege to the philosopher Ludwig Wittgenstein, who was an artillery observer in the Austrian army during the war, have survived. Not surprisingly, they show Frege to have been a patriotic German (Frege, 1976).

[5]Frege (1996).

[6]Sluga (1993); Frege (1976, pp. 8–9).

[7]For the quoted comment, see van Heijenoort (1967, p. 1). The same source includes an excellent translation of Frege's *Begriffsschrift* with commentary, pp. 1–82. Another translation is in Bynum (1972, pp. 101–166).

[8]The symbols we are using are those in common use today not those used by Frege. Of course the fundamental insight was recognizing what needed to be symbolized rather than what specific symbols were used. Frege's were not widely adopted in part because they presented difficulties for the typesetter but mainly because the notation used by the Italian logician Giuseppe Peano as adapted by Bertrand Russell became much better known.

[9]Frege wrote "...what I wanted to create was not a mere *calculus ratiocinator* but a *lingua charactera* in Leibniz's sense." Quoted in van Heijenoort (1967, p. 2). See also Kluge (1977).

[10]This rule is known as *modus ponens*. The terminology derives from the scholastic logicians of the twelfth century.

[11]What we are calling Frege's logic is usually called *first-order logic*. This is to distinguish it from systems of logic in which the quantifiers $\forall$ and $\exists$ are applied to properties as well as to individuals. Here is an example of a sentence in what is known as *second-order logic*:

$$(\forall F)(\forall G)[(\forall x)(F(x) \supset G(x)) \supset (\exists x)(F(x) \supset (\exists x)G(x))].$$

Actually, Frege went beyond first-order logic in that he did consider quantification of properties; so our speaking of first-order logic as "Frege's logic" is not quite accurate.

[12]Strictly speaking, this explanation of "number" is closer to what Bertrand Russell proposed than to Frege's own exposition. But it is close enough to show why it was vulnerable to Russell's paradox.

[13]Interesting work done while this book was being written showed that a considerable part of Frege's program for the logical development of arithmetic can be saved (Boolos, 1995).

[14]Frege (1960).

[15]Dummett (1981); Baker and Hacker (1984).

[16]For clarity it is important to be able to state precisely the meaning of the locutions that occur in computer programming languages, or as one says, to provide the *semantics* of such a language. One approach to this question that has been much studied, known as *denotational semantics,* is ultimately based on Frege's ideas. See Davis et al. (1994, pp. 465–556).

Chapter 4

[1]Rucker (1982, p. 3).

[2]Quoted from Dauben (1979, p.124). This is a translation from Leibniz's original in French (Cantor, 1932, p 179).

[3]Dauben (1979, p. 120).

[4]Frege (1892, p. 272). This citation is of a review by Frege of some of Cantor's work. More will be said about this review later in this chapter.

[5] For biographical information about Cantor, I have relied on Grattan-Guinness (1971), Purkert and Ilgauds (1987), and Meschkowski (1983).

[6] Meschkowski (1983, p. 1) (my translation).

[7] Great interest by mathematicians and physicists in trigonometric series was stimulated by the surprising discovery by the French mathematician Fourier early in the nineteenth century that there was little apparent limitation to what they could converge to. An example of a trigonometric series is

$$\cos x + \frac{\cos 2x}{4} + \frac{\cos 3x}{9} + \frac{\cos 4x}{16} + \frac{\cos 5x}{25} + \cdots.$$

Remarkably, this series converges to $\frac{1}{4}x^2 - \frac{1}{2}\pi x + \frac{1}{6}\pi^2$ if x has any value between 0 and 2π. (The angle x is measured in radians.) If x is set equal to 0, we get

$$\frac{\pi^2}{6} = 1 + \frac{1}{4} + \frac{1}{9} + \frac{1}{16} + \frac{1}{25} + \ldots,$$

a result which, like Leibniz's series for $\frac{\pi}{4}$, connects π with the natural numbers, in this case with the perfect squares:

$$1 \times 1 = 1,\ 2 \times 2 = 4,\ 3 \times 3 = 9,\ 4 \times 4 = 16,\ 5 \times 5 = 25,\ \ldots.$$

[8] Euclid (1956, p. 232).

[9] Gerhardt (1978, v. 1., p. 338). The translation from Latin is by Alexis Manaster Ramer.

[10] As we all learned in elementary school, different fractions can represent the same number, e.g.,

$$\frac{1}{2} = \frac{2}{4} = \frac{3}{6} = \cdots.$$

So the one-to-one match between fractions and natural numbers shown is a match with the *fractions as symbols* rather than with the numbers the symbols stand for. But this is easily fixed: just remove from the list of fractions all those not in lowest terms.

[11] The existence of transcendental numbers had been proved by the French mathematician Liouville in an entirely different manner three decades earlier. What Liouville had been able to prove is that a number whose decimal expansion includes enormously long stretches of 0s had to be transcendental. An example to which Liouville's method would apply is the number

$$.10100001\,\underbrace{000000000000000000000000000}_{27}\,1\,\underbrace{000\ldots0}_{64}\,10\ldots.$$

Here the successive blocks of 0s between the 1s are of lengths $1^1 = 1$, $2^2 = 4$, $3^3 = 27$, $4^4 = 64$, etc. At the time Cantor wrote his paper, a proof that π is transcendental was still a decade away. The fact that $2^{\sqrt{2}}$ is transcendental was not proved until 1934.

[12]Grattan-Guinness (1971, p. 358).

[13]Cantor's notation for cardinal numbers is not used much today. Instead of $\overline{\overline{M}}$, contemporary authors write $|M|$.

[14]In fact, the proposition that of any two unequal cardinal numbers, one must be larger than the other, is not so evident in the case of infinite sets. The matter was not really cleared up during Cantor's lifetime.

[15]To see why the cardinal number of the set of all sets of natural numbers is the same as that of the set of real numbers, it is helpful to consider the representation of numbers using, not the familiar decimal system, but instead, the binary system, in which there are only the two digits 0 and 1. When we write $\frac{1}{3} = .33333\ldots$, that simply means

$$\frac{1}{3} = \frac{3}{10} + \frac{3}{100} + \frac{3}{1000} + \frac{3}{10000} + \cdots.$$

In the binary system, positive real numbers less than 1 are represented by infinite strings of 0s and 1s. For example

$$\begin{aligned} \frac{1}{4} &= .0100000000\ldots, \\ \frac{1}{3} &= .0101010101\ldots, \\ \frac{1}{\pi} &= .0101000101\ldots, \\ \sqrt{\frac{1}{2}} &= .1011010100\ldots. \end{aligned}$$

Here when we write $\frac{1}{3} = .0101010101\ldots$, mean

$$\frac{1}{3} = \frac{1}{4} + \frac{1}{16} + \frac{1}{64} + \cdots.$$

(The denominators are successive powers of 2 instead of 10.)

Now, starting with any set of natural numbers, we can find a unique corresponding real number as follows: we generate a string of 0s and 1s by writing 1 in the nth place if n is a member of the given set and 0 otherwise. For example, if we begin with the set of even numbers, we end up with $.01010101\ldots$, i.e., as we have seen, $\frac{1}{3}$. If instead, we begin with the set of odd numbers, we get

$$.10101010\ldots = \frac{2}{3}.$$

This shows that the set of all sets of natural numbers has the same cardinal number as the set of real numbers between 0 and 1. But Cantor was able to prove (and it's not really difficult) that this set has the same cardinal number as the set of all real numbers.

There is a minor technical nuisance that in good conscience I must mention. Certain rational numbers will have two different binary representations and hence will be matched with two different sets of natural numbers. An example is:

$$\begin{aligned} \frac{1}{2} &= .1000000\ldots \\ &= .0111111\ldots. \end{aligned}$$

So the real number $\frac{1}{2}$ corresponds both to the set consisting only of the number 1 and to the set consisting of all natural numbers except 1. Although this spoils our one-to-one match-up, the difficulty can be overcome using the fact that the set of rational numbers for which this happens has cardinal number $\aleph_0$.

[16]As Cantor pointed out, the cardinal number of the set of all sets of real numbers is also the cardinal number of the set of all *functions* from real numbers to real numbers.

[17]See Grattan-Guinness (1971) and Dauben (1979). Dr. Barbara Rosen kindly provided professional advice to me on this matter.

[18]I'm grateful to Michael Friedman for help with Kant and related matters (although he should not be held responsible for my attack on Hegel).

[19]Quoted from Cantor (1932, pp. 383-384) and Frege (1892). I have profited greatly from the help of Egon Börger, William Craig, Michael Richter, and Wilfried Sieg in translating these, quite difficult for me, passages from German.

I learned about these comments from Dauben (1979, p. 225).

For the benefit of any readers who are fluent in German, I include the original passages. Cantor wrote:

> So sehen wir die in Deutschland als Reaktion gegen den überspannten Kant-Fichte-Hegel-Schellingschen Idealismus eingetretene, jetzt herrschende und mächtige *akademisch-positivistische Skepsis* endlich auch bei der *Arithmetik* angelangt, wo sie, mit der äussersten, für sie selbst vielleicht verhängnisvollsten Konsequenz, die letzten, ihr noch möglichen Folgerung zu ziehen scheint.

Agreeing with Cantor's prediction, Frege wrote:

> In der That! hier ist die Klippe, wo sie schreitern wird. Denn das Unendliche wird sich in der Arithmetik doch schliesslich nicht leugnen lassen, und anderseits ist es mit jene erkentnis-theoretischen Richtung unvereinbar. Hier ist , wie es scheint, das Schlachtfeld, wo eine grosse Entscheidung fallen wird.

[20]Bell (1937, p. 621).

[21]Bell (1986, pp. 562–563). This awkward bowdlerization was almost certainly not written by Bell who had been dead for 26 years when this edition appeared.

[22]Edwards (2009).

[23]Edwards (1988).

[24]Bell (1937, p. 619), Bell (1986, p. 561).

[25]Bell (1937, p. 617), Bell (1986, p. 559).

[26]Bell (1937, p. 629), Bell (1986, p. 570).

[27]For Cantor's letter to Kronecker, see Schönflies (1927, p. 10), and for Kronecker's reply, see Meschkowski (1967, p. 238). Both translations are by Harold Edwards.

[28]Schönflies (1927, pp. 12–13).

[29]Dauben (1979, p. 225). See Note [19] for Cantor and Frege's remarks in the original German.

[30]Dauben (1979, p. 225).

[31]Poincaré (1909, p. 182). The translation given of the sentence beginning "One of the characteristic features of Cantorism" is somewhat anachronistic. A more literal translation is:

> One of the characteristic features of Cantorism is this: instead of achieving generality by building up more and more complicated constructions and defining by construction, it begins with the *genus supremum* and only defines, as the scholastics would have said, *per genus proximum et differentiam specificam.*

I am indebted to Wilfried Sieg and François Treves for help with the translation. I include the original French below:

> J'ai parlé plus haut du besoin que nous avons de remonter sans cesse aux premiers principes de notre science et du profit qu'en peut tirer l'étude de l'esprit humain. C'est ce besoin qui a inspiré deux tentatives qui ont tenu une très grande place dans l'histoire la plus récente des mathématiques. La premiere est le cantorisme, qui a rendu à la science les services que l'on sait. Un des traits caractéristiques du cantorisme, c'est qu'au lieu de s'élever au général en bâtissant des constructions de plus en plus compliquées et de dfinir par construction, il part du *genus supremum* et ne définit, comme auraient dit les scholastiques, que *per genus proximum et differentiam specificam.* De là l'horreur qu'il a quelque temps inspirée à certains esprits, à HERMITE par exemple, dont l'idée favorite était de comparer les sciences mathématiques aux sciences naturelles. Chez la plupart d'entre nous ces préventions s'étaient dissipées, mais il est arrivé qu'on s' est heurté à certains paradoxes, á certaines contradictions apparentes, qui auraient comblé de joie ZÉNON d'Elée et l'école de Mégare. Et alors chacun de chercher le remède. Je pense pour mon compte, et je ne suis pas seul, que l'important c'est de ne jamais introduire que des êtres que l'on puisse définir complètement en un nombre fini de mots. Quel que soit le remède adopte, nous pouvons nous promettre la joie du médecin appelé à suivre un beau cas pathologique.

[32]Dauben (1979, pp. 69–70). Dauben (1995, p. 227).

[33]For example, the section on partial differential equations refers to the so-called "Dirichlet Principle" and the usefulness of Hilbert's efforts to establish it rigorously.

[34]Grattan-Guinness (2000, p. 89).

Chapter 5

[1]For information about Hilbert, I've made use of the biography (Reid, 1986), the biographical essay by Otto Blumenthal (Hilbert, 1935/1970, pp. 388–429), and Hermann Weyl's obituary essay (Weyl, 1944).

[2]Many readers will be familiar with the fact that $\sqrt{2}$ is an irrational number. (As explained in the previous chapter, this means that it cannot be expressed as a fraction with natural numbers as numerator and denominator, or equivalently, that its decimal representation is non-repeating.) Using this fact, it is possible to give an elegant *non-constructive proof* of the following theorem:

There exist irrational numbers a and b such that a^b is rational.

In carrying out the proof, we use the letter q to stand for the number $\sqrt{2}^{\sqrt{2}}$. Now q must be either rational or irrational. If q is rational, we get what we wanted to prove by letting $a = b = \sqrt{2}$. If q is irrational, we can take $a = q$ and $b = \sqrt{2}$. Then,

$$a^b = q^{\sqrt{2}} = \left(\sqrt{2}^{\sqrt{2}}\right)^{\sqrt{2}} = \sqrt{2}^{(\sqrt{2}\cdot\sqrt{2})} = \left(\sqrt{2}\right)^2 = 2;$$

so, once again we have an irrational number raised to an irrational exponent giving a rational number as result. The proof is non-constructive because it doesn't give specific numbers a and b that satisfy the theorem, but only two separate possibilities, one of which must work. (Actually q is irrational, but there is no known easy proof of that fact.)

[3]In the theory of algebraic invariants, it was the so-called *unimodular transformations* that were of particular interest. These took the form of substituting for an unknown quantity (say x) in an equation, the expression $(py+q)/(ry+s)$ where y is a new unknown, and p, q, r, s are particular numbers chosen so that $ps - rq = 1$ or $= -1$. Boole found that for the general quadratic equation $ax^2 + bx + c = 0$ ("general" because the letters a, b, c can stand for any numbers), the expression $b^2 - 4ac$ (which the algebra textbooks call the *discriminant* of the equation) is an *invariant* of such unimodular transformations in the following sense.

After the indicated substitution is made in the given quadratic equation, and after clearing of fractions, a new quadratic equation in the unknown y results. This equation can be written $Ay^2 + By + C = 0$ where A, B, C depend on all the quantities a, b, c, p, q, r, s. The precise sense in which $b^2 - 4ac$ is an invariant is that the new equation has the same discriminant as the given equation, i.e., $b^2 - 4ac = B^2 - 4AC$.

Without the special condition $ps - rq = \pm 1$, the relation between the two discriminants is

$$B^2 - 4AC = (b^2 - 4ac)(ps - rq)^2.$$

Any readers wishing to work this out and are not deterred by a little high-school algebra are advised to begin by writing

$$ax^2 + bx + c = a(x - x_1)(x - x_2)$$

where x_1, x_2 are the two roots of the equation, and by noting that, using the quadratic formula,

$$b^2 - 4ac = 4a^2(x_1 - x_2)^2.$$

[4]In his obituary notice for Hilbert (Weyl, 1944), Hermann Weyl wrote:

> Indeed, by discovering new ideas and introducing new powerful methods he not only brought the subject up to the level set for algebra by Kronecker and Dedekind, but made such a thorough job of it that he all but finished it ... With justifiable pride he concludes his paper, *Über die vollen Invariantensysteme,* with the words: "Thus I believe the most important goals of the theory of ... [algebraic] invariants have been attained," and therewith quits the scene.

[5]In its classical form, the theory of numbers deals with the remarkable relationships and patterns to be found among the natural numbers $1, 2, 3, \ldots$, particularly questions involving prime numbers and divisibility. In algebraic number theory, some of these matters are considered in domains obtained by adjoining to the integers, roots (real or complex) of certain algebraic equations. Gauss had worked with numbers of the form $m + n\sqrt{-1}$ where m, n are ordinary integers, had found which of these "Gaussian integers" were prime, and had proved that the theorem that numbers can be factored into primes in exactly one way, holds for these numbers just as it does for the ordinary integers. However, if one works with numbers having the form $m + n\sqrt{10}$, this turns out not to be the case. A counterexample is

$$6 = 2 \cdot 3 = (2 + \sqrt{10})(-2 + \sqrt{10})$$

where it can be shown that $2, 3, 2 + \sqrt{10}$, and $-2 + \sqrt{10b}$ are all primes so that unique factorization fails. Cantor's friend Dedekind and his nemesis Kronecker had shown how to restore unique factorization by considering what came to be called "prime ideals." On their strolls, Hilbert and his friend Hurwitz, had discussed these competing approaches and agreed that both were *scheusslich* (atrocious). In contrast, the treatment in Hilbert's *Zahlbericht* is elegant.

[6]Hilbert (1935/1970, pp. 400, 401).

[7]There wasn't time during the lecture for Hilbert to state all 23 of his problems, and he contented himself with a selection. For the full address with all 23 problems in an English translation by Mary Winston Newson, see Browder (1976, pp. 1–34).

[8]See Browder (1976). I'm a coauthor of the article on the tenth problem.

[9]The quotation is given in detail at the close of Chapter 4.

[10]van Heijenoort (1967, pp. 129–138).

[11]For Poincaré's criticisms of Cantor, Hilbert, and Russell, see Poincaré (1952, Chapter III).

[12]The technical term for Russell's "elaborate and unwieldy" layers is "the ramified theory of types."

[13]As with Frege's *Begriffsschrift*, the main rule of inference in *Principia* is that which proceeds from a pair of formulas of the form $\mathcal{A} \supset \mathcal{B}$ and $\mathcal{A}$ to the corresponding formula $\mathcal{B}$ (known as *modus ponens* or as the *rule of detachment*). Although Frege is very clear about this, Whitehead and Russell muddy the waters by expressing the rule as their "primitive proposition": *Anything implied by a true proposition is true.* (Whitehead and Russell, 1925, p. 94).

[14]Brouwer (1996).

[15]Brouwer's doctoral dissertation was written in Dutch. An English translation appears in Brouwer (1975, pp. 13–97).

[16]van Stigt (1990, p. 41).

[17]The quote is from Brouwer's dissertation (Brouwer, 1975, p. 96).

[18]In the example given of a non-constructive proof,[2] the law of the excluded middle is used in the assertion "q must be either rational or irrational."

[19]Weyl was upset by the use of so-called *impredicative* definitions in the work of Cantor and Dedekind. Something is defined *impredicatively* if the definition is in terms of a set of which the item being defined is a member. From the point of view of a philosophy in which mathematical objects are "constructed" a bit at a time, such a definition is seen as being objectionable because the set in question cannot have been constructed *before* one of its elements. The contrary philosophical view that mathematical objects are pre-exisiting and definitions merely single them out (like the characterization: Mathilda is the tallest person in the room) rather than construct them is called Platonism and was unacceptable to Weyl.

[20]This was part of an address delivered in 1922, first in Copenhagen and then in Hamburg. I'm indebted to Walter Felscher for calling my attention to the connection between Hilbert's heated rhetoric and the times he was living through. The full text of the address can be found (in English translation) in Mancosu (1998, pp. 198–214). I found the translation accurate enough, but not communicating adequately the fire in the original. In my own attempt to do better, I consulted several translations as well as the original (Hilbert, 1935/1970, pp. 159–160).

[21]Reid (1986, pp. 137–138, 144, 145). For the background of the manifesto by German intellectuals, see Tuchman (1962/1988, p. 322).

[22]Reid (1986, p. 143).

[23]Hilbert (1935/1970, p. 146) (my translation).

[24]Hilbert's program is discussed in an interesting essay in Mancosu (1998, pp. 149–197). See also Sieg (1999) for a thorough discussion and analysis based

on unpublished documents showing clearly the evolution of Hilbert's thought. For interesting information about Bernays's contributions, see Zach (1999). For von Neumann on intuitionism ad absurdum, see Mancosu (1998, p. 168). It should be mentioned that although Hilbert's description of just which methods would be permitted as being "finitary" was never made completely explicit, it is generally agreed that what he had in mind was even more restrictive than what Brouwer was prepared to permit.

[25]van Heijenoort (1967, p. 373).

[26]van Heijenoort (1967, p. 376).

[27]van Heijenoort (1967, p. 336).

[28]Reid (1986, p. 187).

[29]van Stigt (1990, p. 272).

[30]van Stigt (1990, p. 110).

[31]van Stigt (1990, pp. 285–294); Mancosu (1998, pp. 275–285).

[32]Intuitionistic logic in computer science (Constable, 1986).

[33]Hilbert (1935/1970, pp. 378–387).

[34]Dawson (1997, p. 69).

Chapter 6

[1]For Einstein on Gödel voting for Eisenhower, see Dawson (1997, p. 209). I've been fortunate to have this superb biography of Gödel available. I've also made use of the brief collection (Weingartner and Schmetterer, 1983) based on an invitation-only symposium on Gödel in Salzburg in 1983 (that I was privileged to attend). There is much interesting material in the obituary memoir (Kreisel, 1980) by Georg Kreisel, who for a time had been a close friend of Gödel, but unfortunately, it is not entirely reliable. A brief sensitive biography of Gödel by the logician Solomon Feferman is in Gödel (1986/1990, vol.I, pp. 1–36).

[2]Gödel (1986/1990, vol. III, pp. 202–259).

[3]Dawson (1997, pp. 58, 61, 66).

[4]Weingartner and Schmetterer (1983, p. 27).

[5]The phrase "The symbolic logic of Frege-Russell-Hilbert" is an over-simplification. The basic logic that Hilbert singled out, what is known today as first-order logic, was only part of the systems of Frege and of Russell.

[6]For Gödel on the blindness of logicians, see Dawson (1997, p. 58). The complete text of Gödel's dissertation as well as the published article based on it (in the original German as well as in English translation) can be found in Gödel (1986/1990, vol.I, pp. 60–123). An illuminating introductory note by Burton Dreben and Jean van Heijenoort precedes the dissertation and the article and appears on pp. 44–59.

[7]Although Hilbert's finitistic methods in metamathematics are often characterized as "intuitionistic," it is likely that what Hilbert had in mind was even

more restrictive than what Brouwer would permit. For a discussion of this matter, see Mancosu (1998, pp. 167–168).

[8]Gödel (1986/1990, vol. I, p. 65).

[9]Although it is of no real importance, it might be mentioned that the technique Gödel used for coding strings did not use the representation of numbers by decimal digits. Instead he used the fact that factorization of a natural number into prime factors is unique, and placed the code numbers assigned to individual symbols as exponents on the corresponding prime numbers. A simple example should make the difference clear. The string $L(x, y)$ would be coded in our scheme by 186079. In Gödel's scheme the code number would be $2^1 3^8 5^6 7^0 11^7 13^9$.

[10]There have been a number of English translations of this epochal article. The best translation (and the one approved by Gödel) is available both in Gödel (1986/1990, vol. I, pp. 144–145) (page facing with the original German) and in van Heijenoort (1967, pp. 596–616). Readers interested in Gödel's story of how he discovered his incompleteness theorem should see Dawson (1997, p. 61).

[11]To avoid the use of a philosophically suspect notion like "truth," Gödel had recourse to a technical substitute he called omega-consistency, a kind of strengthened consistency property. So the correct statement of his theorem is: *if* PM *is omega-consistent, then there is a proposition U such that neither U nor $\neg U$ is provable in* PM. An important improvement came a few years later when J.B. Rosser showed how to replace the assumption of omega-consistency by that of ordinary consistency. Together with other work that had been done in the meantime (in particular that of Alan Turing to be discussed in the following chapter), it became possible to state Gödel's results in the attractive form: no matter what additional axioms are added to PM, so long as the new axioms are specified by an algorithm and so long as they do not lead to a contradiction (i.e., a proposition of the form $A \wedge \neg A$) being provable, there will be a proposition U undecidable in the system.

[12]After Gödel had proved that the consistency of PM could not be proved using all of the mathematical resources encapsulated in PM, it would have been natural to conclude that it was hopeless to expect success for Hilbert's goal of proving this consistency using the limited finitary methods he was willing to permit. This was certainly von Neumann's conclusion. Gödel was not so sure; the hope he held out was that there might be some proof methods not permitted *inside* PM that could be accepted as "finitary" and which would lead to consistency proofs. What has happened in the decades since Gödel's discovery is that methods have been developed with some claim to meeting this criterion. As a result, Hilbertian proof theory continues to undergo vigorous development as a research area, although few would claim that the consistency theorems that have been proved have added any confidence in the validity of the systems in question.

[13]The programming languages that are mainly in use in the software industry (like C and FORTRAN) are usually described as *imperative*. This is because the successive lines of programs written in these languages can be thought of

as *commands* to be executed by the computer. Object-oriented languages like C++ are also imperative. In the so-called *functional* programming languages (like LISP), the lines of a program are definitions of operations. Rather than telling the computer what to do, they *define* what the computer is to provide. Gödel's special language is very much like a functional programming language.

[14]Returning to the example of PA with the specific encoding we had suggested, we can examine some of the issues involved in translating metamathematical concepts into numerical operations. The first question we can raise is: given the code number for some string, how can we tell how long the string is? Now, since we allowed two digits per symbol, the answer is simple: the length is half the number of digits in the code. For a code number r, let's write $\mathcal{L}(r)$ for the length of the corresponding string. Next: given two strings, a new string can always be formed by placing the second immediately after the first; what is the code of this new string given that the strings have codes r and s, respectively? The answer is given by the formula $r10^{2\mathcal{L}(s)} + s$. This is because multiplying r by this power of 10 has the effect of placing just as many 0s after it as there are digits in s. Following Gödel, we write this $r * s$. Now suppose that r and s are the codes of two sentences; what is the code of the new sentence we get by placing the symbol $\supset$ between them and parentheses around the result? Consulting the coding table we see that the answer is $41 * r * 10 * s * 42$. Continuing in this way, ever more complex metamathematical notions translate into arithmetic operations

[15]The Chinese remainder theorem apparently goes back to the eleventh century in China. The theorem can be illustrated by the following exercise. *Find a number which when divided by* 6 *will leave a remainder of* 2 *and when divided by* 11 *will leave a remainder of* 5. A little experimenting shows that 38 does the job. The Chinese remainder theorem guarantees that a number can always be found leaving given numbers as remainders when divided by other given numbers, so long as no two of these other given numbers have any common factor (except of course 1). So, for example, there will be a number whose remainders on dividing by $3, 7, 10, 11$ are $1, 4, 8, 9$ respectively. But the conclusion cannot be guaranteed if 7 is replaced by 14 (because then the divisors 14 and 10 would have the factor 2 in common). Gödel used the Chinese Remainder Theorem as a coding device: a long sequence of numbers can be specified by a collection of divisors designed to have no pair with a common factor and a single number to be divided by each of them. Since "remainder" is easily definable in the basic language of arithmetic, this could be used to express relationships involving sequences of natural numbers in this language.

Gödel's technique for using the Chinese remainder theorem to code finite sequences of natural numbers played an important role in my own professional life. As part of the research for my doctoral dissertation (accepted by Princeton University in 1950) I worked on the tenth problem in Hilbert's 1900 list, and the Chinese remainder theorem was extremely important for the partial results I was able to obtain. Later work with Hilary Putnam and with Julia Robinson continued to make essential use of this theorem. The crucial final step in the solution of Hilbert's tenth problem was provided by the 22-year-old Russian mathematician Yuri Matiyasevich in 1970. Interested readers can consult the article Davis and Hersh (1973) intended for a general audience.

[16]The full text of the Königsberg addresses by Carnap, Heyting, and von Neumann can be found in Benacerraf and Putnam (1984, pp. 41–65).

[17]For the complete statements of Gödel's remarks at the Königsberg roundtable (in the original German as well as English translations) together with illuminating comments by John Dawson, see Gödel (1986/1990, vol.I, pp. 196–203). See also Dawson (1997, pp. 68–71).

[18]Dawson (1997, p. 70).

[19]Goldstine (1972, p. 174).

[20]This research involves very large transfinite cardinal numbers and is well beyond the scope of this book. For an interesting article by a leading skeptic, see Feferman (1999).

[21]Dawson (1997, pp. 32–33, p. 277).

[22]Dawson (1997, p. 34).

[23]Dawson (1997, p. 111).

[24]Weingartner and Schmetterer (1983, p. 27).

[25]The most interesting of these contributions had to do with certain formal systems developed by Brouwer's student, Heyting, that were intended to encapsulate Brouwer's foundational ideas. Brouwer remained convinced that no precisely defined formal language could do justice to his concepts, but he did express a grudging interest in what Heyting had done. One of Heyting's systems, HA (for Heyting arithmetic) is very much like PA except that for the underlying logic, rules in keeping with what Brouwer thought acceptable are used instead of Frege's rules. In particular, the law of the excluded middle is not available in HA. What Gödel found was a simple way of translating PA into HA, so that, contrary to the idea that intuitionism is narrower than classical mathematics, in this case there is a sense in which it includes it. In particular, any proof of the consistency of HA translates at once into a proof of the consistency of PA.

[26]In the λ-calculus, algorithms use operations of a specified kind. A simple example of such an operation begins with $\{\lambda x[a(x)\}b(c)$ and produces the result $a(b(c))$. This is in accord with the intuition that $\lambda x[a(x)$ denotes a function that replaces x with a given input, in this case $b(c)$. Finally, a function of natural numbers is called λ-*definable* if its values, expressed in the λ-calculus notation for natural numbers, can be calculated by an algorithm consisting of a sequence of such operations.

[27]It is convenient to use a the symbol $\diamond$ as a superscript to represent the successor of a given natural number, i.e., the number that immediately follows it (so, for example, $1^\diamond = 2$ and $4^\diamond = 5$). A *recursive* definition of addition of natural numbers needs to show two things: first, the result of adding 1 to a given number and, second, the result of adding the successor of a given number to another

number. This is accomplished by the equations:

$$x + 1 = x^{\diamond}, \quad \text{and} \quad x + y^{\diamond} = (x + y)^{\diamond}.$$

These equation can be used to calculate the sum of any pair of natural numbers. For example,

$$3 + 2 = 3 + 1^{\diamond} = (3 + 1)^{\diamond} = (3^{\diamond})^{\diamond} = 4^{\diamond} = 5$$

Next we consider the equations

$$x \times 1 = x, \quad \text{and} \quad x \times y^{\diamond} = x \times y + x$$

which furnish a recursive definition of multiplication. This definition together with the preceding definition of addition can be used to calculate the result of multiplying two natural numbers. For example,

$$2 \times 2 = 2 \times 1^{\diamond} = (2 \times 1) + 2 = 2 + 2 = 2 + 1^{\diamond} = (2 + 1)^{\diamond} = (2^{\diamond})^{\diamond} = 3^{\diamond} = 4.$$

The functions in Gödel's original class of recursive functions (renamed primitive recursive functions by Kleene) were built by a succession of such recursive definitions.

Wilhelm Ackermann, one of Hilbert's students, showed that by working with a recursive definition that increased the values of two variables simultaneously, one could define a function that is not primitive recursive. A simpler example of such a "double recursion", was found by the Hungarian mathematician Rósza Péter, and is often incorrectly called the Ackermann function. Both Ackermann's original example and Péter's are general recursive functions that are not primitive recursive. Péter's recursion begins with 0 rather than 1; her equations are:

$$\begin{aligned} g(0, y) &= y + 1 \\ g(x^{\diamond}, 0) &= g(x, 1) \\ g(x^{\diamond}, y^{\diamond}) &= g(x, g(x^{\diamond}, y)). \end{aligned}$$

As an example, we calculate $g(1, 2)$:

$$\begin{aligned} g(1, 2) &= g(0, g(1, 1)) \\ &= g(0, g(0, g(1, 0)))) \\ &= g(0, g(0, g(0, 1)))) \\ &= g(0, g(0, 2)) \\ &= g(0, 3) \\ &= 4. \end{aligned}$$

This function grows very rapidly, faster than any primitive recursive function such as $x^{x^{x^x}}$. Already $g(4, 3) = 2^{65536} - 3$, a number much larger than the total number of atoms in the entire observable universe. General recursive definitions can be far more complicated than such a double recursion. Nevertheless, Kleene was able to prove: *For every general recursive function* $f(x_1, x_2, \ldots, x_n)$, *there are primitive recursive functions* $g(x)$ *and* $h(x_1, x_2, \ldots, x_n, y)$ *such that*

$$f(x_1, x_2, \ldots, x_n) = g(\min_y[h(x_1, x_2, \ldots, x_n, y) = 0]).$$

Here, as the notation suggests, $\min_y[h(x_1, x_2, \ldots, x_n, y) = 0]$ is the least value of y corresponding to given values of $x_1, x_2, \ldots, x_n$ for which $h(x_1, x_2, \ldots, x_n, y) = 0$.

[28]Church (1936).

[29]Dawson (1997, pp. 103–106).

[30]Weingartner and Schmetterer (1983, p. 20).

[31]Dawson (1997, p. 142, 146).

[32]Dawson (1997, p. 91).

[33]Dawson (1997, p. 147).

[34]Dawson (1997, pp. 143–145, 148–151).

[35]Dawson (1997, p. 153).

[36]Browder (1976, p. 8).

[37]More precisely, what Gödel showed is that *if* systems like PM or those based on axioms for set theory are consistent, then they remain consistent if the continuum hypothesis is adjoined as a new axiom. So if these systems are consistent, the continuum hypothesis cannot be disproved in them.

[38]The battle rages on. That the continuum hypothesis is "inherently vague" was the position taken by the eminent logician Solomon Feferman, in Feferman (1999). After some initial wavering, Gödel eventually came to believe that the continuum hypothesis is not at all vague, that in fact, it is a perfectly meaningful assertion, and that most likely it is false.

[39]Gödel (1986/1990, vol. II, pp. 108, 186).

[40]Gödel (1986/1990, vol. III, pp. 49–50).

[41]Gödel (1986/1990, vol. II, pp. 140–141).

[42]Gödel (1986/1990, vol. III) contains most of the previously unpublished works of Gödel.

[43]Dauben (1995, p. 458), for Gödel's hope that Robinson would be his successor; Dauben (1995, pp. 485–486) for the quoted letter.

[44]Dawson (1997, pp. 153, 158, 179–180, 245–253).

Chapter 7

[1]Huskey (1980, p. 300).

[2]Ceruzzi (1983, p. 43).

[3]I have been fortunate to have available Andrew Hodges' poignant, beautifully written biography of Turing (Hodges, 1983).

[4]Hodges (1983, p. 29).

[5]Turing expressed his feelings about his dead friend in vivid terms: Alan "worshiped the ground he trod on" and he "made everyone else seem so ordinary" (Hodges, 1983, p. 35, p. 53).

[6]Hodges (1983, p. 57).

[7]Hodges (1983, p. 94).

[8]Actually, Hilbert did not put the *Entscheidungsproblem* in quite that way: he asked for a procedure to determine whether a given expression of first order logic is valid in every possible interpretation. However, after Gödel had proved his completeness theorem, it became clear that the form in which the problem is stated here is equivalent to Hilbert's formulation.

[9]Work on the *Entscheidungsproblem* mainly dealt with expressions called *prenex formulas.* These are expressions involving the logical symbols $\neg \quad \supset \quad \wedge \quad \vee \quad \exists \; \forall$ with the property that all occurrences of the so-called existential and universal quantifiers, $(\exists ..)$ $(\forall ..)$, are at the beginning of the expression (reading left to right) preceding all other symbols. It was not difficult to prove that the *Entscheidungsproblem* could be reduced to the problem of providing an algorithm for determining for a given prenex formula, whether it is *satisfiable*, that is, whether there is some way of interpreting the non-logical symbols in the formula so that it expresses a *true* sentence. To illustrate this concept, consider the two prenex formulas:

$$(\forall x)(\exists y)(r(x) \supset s(x,y)) \text{ and } (\forall x)(\exists y)(q(x) \wedge \neg q(y)).$$

The first is satisfiable: for example, we can take the variables x, y to stand for people alive at some particular moment, we can interpret $r(x)$ to mean "x is a monogamously married man" and $s(x, y)$ to mean "y is the wife of x"; so, with this interpretation, the first prenex formulas says simply "every monogamously married man has a wife," certainly a true statement. On the other hand, the second prenex formula is not satisfiable because no matter what universe of individuals is selected, and no matter how the symbol q is interpreted, this formula would stipulate that all individuals have the property that q represents and that some individual does not.

Prenex formulas can be classified by the particular pattern of existential and universal quantifiers with which they begin. Thus, for example, one speaks of the *prefix class* $\forall\exists\forall$ to mean the set of all prenex formulas beginning $(\forall ..)(\exists ..)(\forall ..)$, and so on. In a paper published by Kurt Gödel in 1932, he produced an algorithm that could test for satisfiability any prenex formula belonging to the prefix class

$$\forall\forall\exists \ldots \exists.$$

In a paper published a year later, he proved that to solve the *Entscheidungsproblem*, it would suffice to provide an algorithm to test the satisfiability of all prenex formulas in the prefix class

$$\forall\forall\forall\exists \ldots \exists.$$

Thus, the gap between what had been done and what was needed had been reduced to a universal quantifier, a single $\forall$.

The relevant papers by Gödel (in the original German as well as in English translation) will be found in Gödel (1986/1990, vol. I, pp. 230–235, 306–327). An illuminating introduction by Warren Goldfarb in the same volume, pp. 226–231, describes some of the earlier work on the problem as well.

[10]Hodges (1983, p. 93).

[11]Turing's discussion of this point was more careful (Turing, 1936, pp. 250–251). (Reprinted: Davis (1965, pp. 136–137); Turing (2001, pp. 18–19); Copeland (2004, pp. 75–77).)

[12]Although the unsolvability of the *Entscheidungsproblem* could be proved in the manner described it would be pretty messy because of the need to develop Turing machine structures for handling integers written in decimal notation. To approach what Turing actually did, we first show that the problem of determining of a given Turing machine whether it will ever halt when started with a totally blank tape, is unsolvable. For suppose there was an algorithm for this problem. Then here is an algorithm for testing membership in D: To test whether a code number n belongs to D, we first write the quintuples making up the Turing machine $\mathcal{T}$ with code number n. Then we write quintuples that cause that n to be written on a Turing machine tape. Adjoining those quintuples to those of the machine $\mathcal{T}$, we get a new machine that will first put n on its tape, and then do what $\mathcal{T}$ would have done with that input. This new machine will eventually halt when started with a blank tape if and only if $\mathcal{T}$ will eventually halt when started with n on its tape, which in turn is true if and only if n doesn't belong to D. So a supposed algorithm for testing whether a given Turing machine started on a blank tape will eventually halt could be used to solve the unsolvable problem of determining membership in D.

Next we notice that the problem of finding out whether a given Turing machine ever prints one particular symbol is also unsolvable. This is because it is easy to arrange matters so that whenever a Turing machine halts it finds itself in a state F which begins no quintuples. We choose a new symbol X that doesn't occur in any of the quintuples of the machine. We then adjoin the quintuples:

$$\mathrm{F}\ a : \mathrm{X} \star \mathrm{F}$$

where a can be any of the symbols that occur in the original quintuples. This new machine will then print X whenever the original machine would have halted. Thus we have that there is no algorithm to determine whether a Turing machine starting with a blank tape will ever print some particular symbol. This is the problem that Turing expressed in the language of first order logic and thus obtained the unsolvability of the *Entscheidungsproblem.*

[13]Turing (1936, pp. 243–246). (Reprinted: Davis (1965, pp. 129–132); Turing (2001, pp. 31–34); Copeland (2004, pp. 69–72).)

[14]Davis (1965, pp. 71–72), Davis (1982).

[15]For a reprint of Turing's dissertation, see Davis (1965, pp. 155-222). It may be mentioned that the hierarchies mentioned extended into Cantor's transfinite.

So, after a 1st, 2nd, 3rd, etc. system, would come system ω, followed by system $\omega + 1$, etc.

[16]Hodges (1983, p. 131).

[17]Hodges (1983, p. 124).

[18]Hodges (1983, p. 145). To those familiar with the later work of Kolmogorov and Chaitin on descriptive complexity, this game may well suggest that von Neumann was thinking along those lines.

[19]Hodges (1983, p. 545).

[20]The anecdote about Turing's adventures with the Home Guard was recounted by the mathematician Peter Hilton, a co-worker with Turing at Bletchley Park (Hodges, 1983, p. 232).

[21]This work was by no means a solo undertaking. Probably the person who made the greatest contribution was W. T. Tutte. For a technical description of the issues addressed by Professor Tutte including the part played by Turing see the web site: http://home.cern.ch/~frode/crypto/tutte.html

Chapter 8

[1]See also M. Davis and V. Davis (2005).

[2]For this quote, see Goldstine (1972, p. 22). The fascinating biography of Ada Lovelace (Stein, 1987) suggests that much that has been written about her is myth rather than fact. See also M. Davis and V. Davis (2005).

[3]Goldstine (1972, p. 120).

[4]Atanasoff's machine was designed to solve simultaneous systems of linear equations. An example of this kind of problem is:

$$\begin{aligned} 2x + 3y - 4z &= 5, \\ 3x - 4y + 2z &= 2, \\ x - 3y - 5z &= 4. \end{aligned}$$

The machine was designed to handle as many as 30 equations in 30 unknowns.

[5]Lee (1995, p. 44). The biographical material in this section is largely derived from this source.

[6]A. Burks and A. Burks (1981).

[7]Differential analyzers contained a number of modules designed to calculate suitable numerical approximations to the value of definite integrals. The ENIAC contained modules that did the same thing, but more accurately, using well-known algorithms for this purpose.

[8]Goldstine (1972, p. 186, p. 188).

[9]Although von Neumann's *First Draft of a Report on the EDVAC* was widely circulated and was very influential, it was only published in 1981 as an appendix to a book rather skeptical about the significance of his contributions (Stern (1981, pp. 177–246)). See also Dyson (2012).

[10]McCulloch and Pitts (1945/1965); Von Neumann (1963, p. 319).

[11]Goldstine (1972, p. 191).

[12]Randell (1982, p. 384).

[13]Goldstine (1972, p. 209); Knuth (1970).

[14]Von Neumann (1963, pp. 1–32).

[15]Von Neumann (1963, pp. 34–79).

[16]For examples of studies that minimize von Neumann's contributions to the development of computers and ignore Turing's entirely, see Metropolis and Worlton (1980) and Stern (1981). For excerpts from Eckert's memo (it is an engineer's "disclosure"), see Stern (1981, p. 28).

[17]Stern (1981) discusses the vicissitudes of the Eckert-Mauchly commercial endeavors.

[18]The analysis of the ACE report quoted is from the excellent paper by Carpenter and Doran (1977). The report itself can be found in Turing (1992, pp. 1–86). For many years it circulated only in mimeographed form and was not easily available.

[19]What Turing proposed was, in contemporary terminology, the use of a *stack* for subroutine management. A *stack* is simply an arrangement of data in a last-in-first-out (LIFO) structure. Thus, when a computation is interrupted to make use of a previously programmed subroutine, a reminder would be noted of what had to be done after the subroutine terminated. Since subroutines could call other subroutines, this would lead to a stack of such reminders. Turing suggested the picturesque terms "bury" for placing a reminder on the stack and "unbury" for retrieving it from the "top" of the stack. (Nowadays the terms PUSH and POP are used.)

[20]Hodges (1983, p. 352).

[21]Turing (1992, pp. 87–88); Copeland (2004, pp. 378–379).

[22]Hodges (1983, p. 361). For Turing's text: Turing (1992, pp. 102–105); Copeland (2004, pp. 392–394).

[23]Metropolis and Worlton (1980); Stern (1981).

[24]Goldstine (1972, pp. 191–192).

[25]Turing (1992, p. 25).

[26]Davis (1988).

[27]Whitemore (1988).

[28]Marcus (1974, pp. 183–184). The book quoted is Engels' famous *The Condition of the Working Class in England in 1844*.

[29]Lavington (1980, pp. 31–47).

[30]Goldstine (1972, p. 218).

[31]Hodges (1983, p. 149).

Chapter 9

[1]Turing (1992, p. 103); Copeland (2004, p. 392).

[2]The five computer scientists who spoke at the AAAS meeting together with the titles of their presentations were as follows:

Joseph Y. Halpern, *Epistemic Logic in Multi-Agent Systems*;

Phokion G. Kolaitis, *Logic in Computer Science: An Overview*;

Christos Papadimitriou, *Complexity As Metaphor*;

Moshe Y. Vardi, *From Boole to the Pentium*;

Victor D. Vianu, *Logic As a Query Language*.

[3]Lee (1995, p. 724).

[4]Turing (1950). Reprinted: Turing (1992, pp. 133–160); Copeland (2004, pp. 433-464).

There has been much discussion concerning the precise setup Turing had in mind, and what the significance would be of a machine passing the "test." I don't propose to say anything about these matters.

[5]Turing (1950, p. 442). Reprinted: Turing (1992, p. 142); Copeland (2004, p. 449).

As Jack Copeland has called to my attention, in previous editions of this book and elsewhere, I have carelessly stated that Turing predicted that more would be achieved by 2000 than what he actually did predict. As to whether the 70% prediction has been met by today's technology, I will not attempt to say. An "average interrogator" can be very gullible. However, I wanted to point out that Turing had not foreseen how difficult it would be to develop a computer program that had the capability with ordinary English conversation of an elementary school child. I do not believe that this has been accomplished, and such conversational ability is surely a necessary condition for even this 70% result to be achieved in any meaningful way. However my making this point hardly makes me one of "Turing's critics" as Copeland asserted in Copeland et al (2017, p. 272). No one familiar with what I have written about Turing and his work could imagine that my remarks were intended as criticism. Given the primitive technology Turing had available, it was only someone with his audacious vision who could at that time have imagined computers as thinking. In fact, I had proclaimed the importance of Turing's ideas at a time when his thought was still quite neglected among computer scientists and historians of computation (Davis, 1988).

[6]The article Searle (1999) contains references to some of his other writing on related topics. The piece is actually a review of a popular book by Ray Kurzweil. It is no part of my purpose to defend Kurzweil, whose ideas have achieved some notoriety, from Searle's onslaught, but only to use the review as a convenient source for some of Searle's often expressed views. Kurzweil has been willing to prophesy a role and capability for computers within a sort time span that most

would consider at least extravagant. He imagines that some kind of symbiotic relation between people and computers will enable a kind of immortality, and he predicts that this will be feasible by 2040.

[7]McCulloch and Pitts (1945/1965).

[8]The function that maps the values arriving at the input channels to a neuron to the total numerical input is linear. Likewise the functions that map data from one portion of a video image to another are linear. So the numerical operations are those students encounter in a course in linear algebra such as matrix multiplication. GPUs are specialized to perform these operations efficiently.

[9]Gödel's theorem could only have been stated in this way after the notion of algorithmic process had been elucidated by Turing, Church, and others.

[10]Penrose first made this case in his popular and entertaining book (Penrose, 1989). Although a number of logicians have tried to set him straight, he continues to hold his misguided views. For an essay that I have written on this subject; see Davis (1990). Penrose (1990) contains replies to his critics, and Davis (1993) is my reply to his replies.

[11]For more information about this and further references, see Gödel (1986/1990, vol. II, p. 297).

Bibliography

Aiton, E.J., *Leibniz: a Biography,* Adam Hilger Ltd., Bristol and Boston, 1985.

Baker, G. P., and P. M. S. Hacker, *Frege: Logical Excavations,* Oxford University Press, New York; Basil Blackwell, Oxford, 1984.

Barret-Ducrocq, Francoise, *Love in the Time of Victoria,* Penguin Books, 1992. Translation by John Howe of *L'Amour sous Victoria,* Plon, Paris, 1989.

Bell, E.T., *Men of Mathematics,* Victor Gallancz London, 1937.

Bell, E.T., *Men of Mathematics,* Touchstone, Simon & Schuster, 1986.

Benacerraf, Paul and Hilary Putnam, (eds.) *Philosophy of Mathematics: Selected Readings,* 2nd Edition, Cambridge University Press, Cambridge, 1984.

Boole, George, *The Mathematical Analysis of Logic, Being an Essay towards a Calculus of Deductive Reasoning,* Macmillan, Barclay and Macmillan, Cambridge, 1847.

Boole, George, *An Investigation of the Laws of Thought on which Are Founded the Mathematical Theories of Logic and Probabilities,* Walton and Maberly, London 1854; reprinted Dover, New York, 1958.

Boole, George, *A Treatise on Differential Equations,* 5th Edition, Macmillan, London, 1865.

Boolos, George, "Frege's Theorem and the Peano Postulates," *The Bulletin of Symbolic Logic,* vol. 1 (1995), pp. 317–326.

Bourbaki, Nicholas, *Eléments d'Histoire des Mathématiques,* Deuxième édition, Hermann, Paris 1969.

The Encyclopedia Britanica, 11th Edition, Cambridge, 1910, 1911.

Brouwer, L. E. J., *Collected Works,* vol. I, edited by A. Heyting. North-Holland, Amsterdam 1975.

Brouwer, L. E. J., "Life, Art, and Mysticism," translated by Walter P. van Stigt, *Notre Dame Journal of Formal Logic,* vol. 37 (1996), pp. 389–429. Introduction by the translator, *ibid,* pp. 381–387.

Browder, Felix (ed.), "Mathematical Developments Arising from Hilbert's Problems," *Proceedings of Symposia on Pure Mathematics,* vol. XXVIII, American Mathematical Society, Providence, 1976.

Burks, Arthur W. and Alice R. Burks, "The ENIAC: First General-Purpose Electronic Computer," *Annals of the History of Computing*, vol. 3 (1981), pp. 310–399.

Bynum, Terrell Ward (ed. and trans.), *Conceptual Notation and Related Articles* by Gottlob Frege (with a biography, introduction, and bibliography by the editor), Oxford University Press, London, 1972.

Cantor, Georg, *Gesammelte Abhandlungen,* Ernst Zermelo, (ed.), Julius Springer, Berlin, 1932.

Cantor, Georg, *Contributions to the Founding of the Theory of Transfinite Numbers,* translated from the German with an introduction and notes by Philip E. B. Jourdain, Open Court, La Salle, IL, 1941.

Carpenter, B. E. and R. W. Doran, "The Other Turing Machine," *Computer Journal*, vol. 20 (1977), pp. 269–279.

Carroll, Lewis (pseud.), *Sylvie and Bruno,* MacMillan and Co., London 1890. Reprinted with an introduction by Martin Gardner, Dover Publications, New York, 1988.

Ceruzzi, Paul E., *Reckoners, the Prehistory of the Digital Computer, from Relays to the Stored Program Concept, 1933–1945*, Greenwood Press, Westport, CT, 1983.

Church, Alonzo, "An Unsovable Problem of Elementary Number Theory," *American Journal of Mathematics,* vol. 58, pp. 345-363. Reprinted: Davis (1965) pp. 89-107.

Constable, Robert L. et al., *Implementing Mathematics with the Nuprl Proof Development System,* Prentice-Hall, Englewood Cliffs, NJ, 1986.

Copeland, B. Jack (ed.), *The Essential Turing,* Oxford University Press, New York, 2004.

Copeland, B. Jack et al, *The Turing Guide,* Oxford, 2017.

Couturat, Louis, *La Logique de Leibniz d'Après des Documents Inédits,* F. Alcan, Paris, 1901. Reprinted: Georg Olms, Hildesheim, 1961.

Craig, Gordon A., *Germany 1866–1945,* Oxford University Press, Oxford, 1978.

Daly, Douglas C., "The Leaf that Launched a Thousand Ships," *Natural History,* vol. 105 no. 1 (January 1996), pp. 24–32.

Dauben, Joseph Warren, *Georg Cantor: His Mathematics and Philosophy of the Infinite,* Princeton University Press, Princeton, NJ, 1979.

Dauben, Joseph Warren, *Abraham Robinson: The Creation of Nonstandard Analysis, a Personal and Mathematical Odyssey,* Princeton University Press, Princeton, NJ, 1995.

Dauben, Joseph Warren, "The Battle for Cantorian Set Theory," in G. Van Brummelen and M. Kinyon, eds., *Mathematics and the Historian's Craft,* pp. 221–241, Springer, New York, 2005.

Davis, Martin (ed.), *The Undecidable,* Raven Press, New York, 1965. Reprinted: Dover, New York, 2004.

Davis, Martin, "Why Gödel Didn't Have Church's Thesis," *Information and Control* vol. 54 (1982), pp. 3–24.

Davis, Martin, "Mathematical Logic and the Origin of Modern Computers," in *Studies in the History of Mathematics,* pp. 137–165, Mathematical Association of America, Washington, DC, 1987. Reprinted in *The Universal Turing Machine — A Half-Century Survey,* Rolf Herken (ed.), pp. 149–174, Verlag Kemmerer & Unverzagt, Hamburg, 1988; Oxford University Press, New York, 1988.

Davis, Martin, "Is Mathematical Insight Algorithmic?" *Behavioral and Brain Sciences,* vol. 13 (1990), pp. 659–660.

Davis, Martin, "How Subtle is Gödel's Theorem? More on Roger Penrose," *Behavioral and Brain Sciences,* vol. 16 (1993), pp. 611–612.

Davis, Martin and Virginia Davis, "Mistaken Ancestry: The Jacquard and the Computer," *Textile,* vol. 3 (2005), pp. 76–87.

Davis, Martin and Reuben Hersh, "Nonstandard Analysis," *Scientific American,* vol. 226 (1972), pp. 78–86.

Davis, Martin and Reuben Hersh, "Hilbert's 10th Problem," *Scientific American,* vol. 229 (1973), pp. 84–91.

Davis, Martin, Ron Sigal, and Elaine Weyuker, *Computability, Complexity, and Languages,* 2nd Edition, Academic Press, New York, 1994.

Dawson, John W., Jr., *Logical Dilemmas: The Life and Work of Kurt Gödel,* A K Peters, Wellesley, MA, 1997.

Dummett, Michael, *Frege: Philosophy of Language,* 2nd Edition, Harvard University Press, Cambridge, MA, 1981.

Dyson, George, *Turing's Cathedral: The Origins of the Digital Universe,* Pantheon, New York, 2012.

Edwards, Charles Henry, Jr., *The Historical Development of the Calculus,* Springer, New York, 1979.

Edwards, Harold, *Kronecker's Place in History,* in: "History and Philosophy of Modern Mathematics," W. Aspray and P. Kitcher, eds., *Minnesota Studies in the Philosophy of Science,* vol. 11, University of Minnesota Press, Minneapolis, 1988.

Leibniz, Gottfried W., "Machina arithmetica in qua non additio tantum et subtractio set et multiplicato nullo, divisio vero pæne nullo animi labore peragantur," 1685. English translation by Mark Kormes in David Eugene Smith, *A Source Book in Mathematics*, pp. 173–181, McGraw-Hill, New York, 1929.

Lewis, C. I., *A Survey of Symbolic Logic,* Dover, New York, 1960. (Corrected version of Chapters I–IV of the original edition, University of California Press, Berkeley, CA, 1918.)

MacHale, Desmond, *George Boole: His Life and Work,* Boole Press, Dublin, 1985.

Mancosu, Paolo, *From Brouwer to Hilbert,* Oxford University Press, New York, 1998.

Mates, Benson, *The Philosophy of Leibniz: Metaphysics & Language,* Oxford University Press, New York, 1986.

McCulloch, W. S. and W. Pitts, "A Logical Calculus of the Ideas Immanent in Nervous Activity," *Bulletin of Mathematical Biophysics*, 5(1943), 115–133. Reprinted in McCulloch, W. S., *Embodiments of Mind*, pp. 19–39, MIT Press, Cambridge, MA, 1965.

Marcus, Steven, *Engels, Manchester, and the Working Class* W. W. Norton, New York, 1974.

Meschkowski, Herbert, *Probleme des Unendlichen,* Vierweg, Braunschweig 1967.

Meschkowski, Herbert, *Georg Cantor: Leben, Werk und Wirkung,* Bibliographisches Institut, Mannheim, 1983.

Metropolis, N. and J. Worlton, "A Trilogy of Errors in the History of Computing," *Annals of the History of Computing*, vol. 2 (1980), pp. 49–59.

von Neumann, John, *First Draft of a Report on the EDVAC*, Moore School of Electrical Engineering, University of Pennsylvania, Philadelphia, 1945. First printed in Stern (1981), pp. 177–246.

von Neumann, John, *Collected Works*, vol. 5, A.H. Taub (ed.), Pergamon Press, New York, 1963.

Parkinson, G. H. R., *Leibniz—Logical Papers*, Oxford University Press, New York, 1966.

Penrose, Roger, *The Emperor's New Mind,* Oxford University Press, London, 1989.

Penrose, Roger, "The Nonalgorithmic Mind," *Behavioral and Brain Sciences,* vol. 13 (1990), pp. 692–705.

Petzold,Charles, *The Annotated Turing: A Guided Tour through Alan Turing's Historic Paper on Computability and the Turing Machine,* Wiley, New York, 2008.

Poincaré, Henri, "L'avenir des Mathématiques," *Atti del IV Congresso Internazionale dei Matematici,* Rome 1909. pp. 168–182. Reprinted: Kraus, Nendeln, Liechtenstein.

Poincaré, Henri, *Science and Method,* Dover, New York, 1952.

Purkert, Walter, and Hans Joachim Ilgauds, in *Georg Cantor: 1845–1918*, Vita mathematica, vol. 1, Birkhauser, Stuttgart, 1987.

Randell, Brian (ed.), *The Origins of Digital Computers, Selected Papers*, 3rd Edition, Springer, New York, 1982.

Reid, Constance, *Hilbert–Courant,* Springer, New York, 1986. (Originally published by Springer as two works: *Hilbert*, 1970 and *Courant in Göttingen and New York: The Story of an Improbable Mathematician*, 1976.)

Rucker, Rudy, *Infinity and the Mind: The Science and Philosophy of the Infinite,* Birkhauser, Boston, 1982.

Schönflies, Arthur, "Die Krisis in Cantor's mathematischen Schaffen," *Acta Mathematica,* vol. 50, pp. 1–23.

Searle, John R., "I Married a Computer," *The New York Review of Books,* April 8, 1999, pp. 34–38.

Sieg, Wilfried, "Hilbert's Programs: 1917–1922" *Bulletin of the Association for Symbolic Logic,* vol. 5 (1999), pp. 1–44.

Siekmann, Jörg and Graham Wrightson (eds.), *Automation of Reasoning*, vol. 1, Springer, New York 1983.

Sluga, Hans, *Heidegger's Crisis: Philosophy and Politics in Nazi Germany,* Harvard University Press, Cambridge, 1993.

Stein, Doris, *Ada: A Life and a Legacy,* MIT Press, Cambridge, 1987.

Stern, Nancy, *From Eniac to Univac: An Appraisal of the Eckert-Mauchly Machines*, Digital Press, Bedford, MA, 1981.

van Stigt, Walter P., *Brouwer's Intuitionism*, North-Holland, Amsterdam, 1990.

Swoyer, Chris, "Leibniz's Calculus of Real Addition," *Studia Leibnitiana*, vol. XXVI (1994), pp. 1–30.

Tuchman, Barbara W., *The Guns of August,* Macmillan, New York, 1962, 1988.

Turing, Alan, "On Computable Numbers with an Application to the Entscheidungsproblem," in *Proceedings of the London Mathematical Society*, Ser. 2, 42 (1936), pp. 230–267. Correction: ibid, 43 (1937), pp. 544–546. Reprinted in Davis (1965) pp. 116–154. Reprinted in Turing (2001) pp. 18–56. Reprinted in Copeland (2004) pp. 58–90; 94–96. Reprinted in Petzold (2008) (the original text interspersed with commentary).

Turing, Alan, "Computing Machinery and Intelligence," *Mind*, vol. LIX, (1950), pp. 433–460. Reprinted in Turing (1992) pp. 133–160. Reprinted in Copeland (2004) pp. 433–464.

Turing, Alan *Collected Works: Mechanical Intelligence*, D.C. Ince (ed.), North-Holland, Amsterdam, 1992.

Turing, Alan *Collected Works: Mathematical Logic*, R.O Gandy and C.E.M. Yates (eds.), North-Holland, Amsterdam, 2001.

Welchman, Gordon, *The Hut Six Story*, McGraw-Hill, New York, 1982.

Weyl, Hermann, "David Hilbert and His Mathematical Work," *Bulletin of the American Mathematical Society,* vol. 50 (1944). pp. 612–654.

Whitehead, Alfred North and Bertrand Russell, *Principia Mathematica,* vol. I, 2nd Edition, Cambridge University Press, Cambridge, 1925.

Whitemore, Hugh, *Breaking the Code,* Samuel French Ltd., London, 1988.

Zach, Richard, "Completeness before Post: Bernays, Hilbert, and the Development of Propositional Logic," *Bulletin of Symbolic Logic,* vol. 5 (1999), pp. 331–366.

Index